Handling, Transportation and Storage of Fruits and Vegetables

Dr. S.K. Chattopadhyay a renowned authority on postharvest technology and corporate farming, has had a brilliant academic record. He had worked with various institutes and has participated in many National and International Conferences. He had presented many papers of international repute in national as well as international seminars. As prolific writer, he authored a number of books on agricultural science and technology.

Handling, Transportation and Storage of Fruits and Vegetables

Editor
Dr. S.K. Chattopadhyay

2025

Gene-Tech Books
New Delhi - 110 002

ISBN : 978-81-89729-74-5

Published by : **GENE-TECH BOOKS**
4762-63/23, Ansari Road, Darya Ganj,
NEW DELHI - 110 002
Phone: 011 - 43003222
e-mail: genetechbooks@yahoo.co.in

Printed in india

Preface

This book has been written with the purpose of covering all biological and physical aspects of the post harvest physiology of fruits and vegetables.

Assembling a substantial amount of information from a wide variety of sources, the book is a compilation of knowledge pertaining to post harvest technology of fruits and vegetables. It expands, in comprehensive detail, on technology relating to maturity determination, harvesting, packing, post harvest treatments, controlled atmosphere storage, ripening and transportation. It also provides meticulous information on storage atmosphere or time constraints affecting acidity, taste, or firmness / texture qualities, optimum refrigeration temperatures and safety concerns of many fruits and vegetables.

With pertinent information on such a matter of great relevance and importance, it is hoped that this book would be useful for researchers and students in food science, food technology, plant and agricultural science.

Editor

Contents

	Preface	*v*
1.	Handling of Fruits and Vegetables Harvesting	1
2.	Microbial Hazards in Fruits and Vegetables	38
3.	Transportation of Fruits and Vegetables	79
4.	Preservation of Fruits and Vegetables	108
5.	Drying Fruits and Vegetables	147
6.	Storage of Fruits and Vegetables	164
7.	Quality in Fruits and Vegetables	203
8.	Hygiene and Sanitation	227
9.	Selling Fruits and Vegetables	242
10.	Preparation for the Fresh Market	258
	Bibliography	277
	Index	279

1

Handling of Fruits and Vegetables Harvesting

Modernization in food production and supply systems has led to increased concentration in the food supply chain in many countries, with the rapid rise of supermarkets. The main beneficiaries have adapted by acquiring managerial skills, adopting new technology and making the necessary investments. This process has been gradual and has had an impact on many small farmers. It has led to the current situation whereby both high and low-cost technology farming systems co-exist. Suppliers in many current food distribution systems use high technology. On the other hand, there has been increased poverty and marginalization of many small farmers.

Mechanical damage during harvest can become a serious problem, as injuries predispose produce to decay, increased water loss and increased respiratory and ethylene production rates leading to quick deterioration. In general, harvesting by machine will cause more damage than harvesting by hand, although some root crops can be severely damaged by careless hand digging. The containers used by pickers in the field should be clean, have smooth inside surfaces and be free of rough edges.

Stackable plastic crates, while initially expensive, are durable, reusable and easily cleaned. If baskets are used, they should be woven "inside out" with the stubs of the beginning and end of each cane on the outside of the basket. Manual harvesters should be well trained in the proper way to harvest the crop to minimise damage and waste, and should be able to recognise the proper maturity stage for the produce they are handling.

Pickers should harvest with care, by snapping, cutting or pulling the fruit or vegetable from the plant in the least damaging manner. The tips of knives should be rounded to minimise inadvertent gouges and excess damage to perennial plants. Knives and clippers should always be well sharpened.

Pickers should be trained to empty their picking bags and/or baskets with care, never dumping or throwing produce into field containers. If harvesters pick directly into large bulk bins, produce can be protected from bruising by the use of a deaccelerating chute fashioned from canvas. Vented, stackable field containers should be kept clean and smooth.

Exposure to the sun should be avoided as much as possible during and after harvest, as produce left out in the sun will gain heat and may become sun-burned. Field bins should be placed in the shade or loosely covered (for example with light colored canvas, leafy plant materials, straw or an inverted empty container) if delays are expected in removing them from the field.

Night or early morning harvest is sometimes an option for harvesting produce when internal temperatures are relatively low, reducing the energy needed for subsequent cooling. Latex flow is often lower later in the morning than it is at dawn for crops such as mango and citrus, so harvesting as late in the morning as possible can also reduce later efforts required to clean the produce before packing.

Directly following harvest, when produce is prepared for marketing, cooling is essential. Cooling (also known as "pre-cooling") is the removal of field heat directly after harvest, before any further handling. Any delays in cooling will shorten postharvest life and reduce quality. Even produce undergoing repeated cooling and warming deteriorates at a slower rate than produce that has not been cooled. Rough handling during preparation for market will increase bruising and mechanical damage and limit the benefits of cooling.

Roads between the field and the packinghouse should be graded and free from large ruts, bumps and holes. Field boxes must be well-secured during transport and, if stacked, not overfilled. Transport speeds must be suited to the quality and conditions of the roads, and truck and/or trailer suspensions kept in good repair. Reduced tire air pressure on transport vehicles will reduce the amount of motion transmitted to the produce.

Any practice that reduces the number of times the produce is handled will help reduce losses. Field packing (selection, sorting, trimming and packaging of produce at the time of harvest) can greatly reduce the number of handling steps the produce must undergo before marketing. Small, mobile field packing stations can be designed to be moved along with the packers and to provide shade for packing operations.

Maturity Index

The principles dictating at which stage of maturity a fruit or vegetable should be harvested are crucial to its subsequent storage and marketable life and quality. Post-harvest physiologists distinguish three stages in the life span of fruits and vegetables: maturation, ripening, and senescence. Maturation is indicative of the fruit being ready for harvest. At this point, the edible part of the fruit or vegetable is fully developed in size, although it may not be ready for immediate consumption. The following table provides some examples of maturity indices.

Index	Examples
Elapsed days from full bloom to harvest	Apples, pears
Mean heat units during development	Peas, apples, sweet corn
Development of abscission layer	Some melons, apples, feijoas
Surface morphology and structure	Cuticle formation on grapes, tomatoesNetting of some melonsGloss of some fruits (development of wax)
Size	All fruits and many vegetables
Specific gravity	Cherries, watermelons, potatoes
Shape	Angularity of banana fingersFull cheeks of mangosCompactness of broccoli and cauliflower
Solidity	Lettuce, cabbage, brussels sprouts
Textural properties	
Firmness	Apples, pears, stone fruits
Tenderness	Peas
Color, external	All fruits and most vegetables
Internal color and structure	Formation of jelly-like material in tomato fruitsFlesh color of some fruits
Compositional factors	
Starch content	Apples, pears
Sugar content	Apples, pears, stone fruits, grapes

Acid content, sugar/acid ratio	Pomegranates, citrus, papaya, melons, kiwifruit
Juice content	Citrus fruits
Oil content	Avocados
Astringency (tannin content)	Persimmons, dates
Internal ethylene concentration	Apples, pears

Vegetables are harvested over a wide range of maturities, depending upon the part of the plant used as food. The following table provides some examples of maturity indices of vegetable crops.

Crop	Index
Root, bulb and tuber crops	
Radish and carrot	Large enough and crispy (overmature if pithy)
Potato, onion, and garlic	Tops beginning to dry out and topple down
Yam bean and ginger	Large enough (overmature if tough and fibrous)
Green onion	Leaves at their broadest and longest
Fruit vegetables	
Cowpea, yard-long bean, snap bean, batao, sweet pea, and winged bean	Well-filled pods that snap readily
Lima bean and pigeon pea	Well-filled pods that are beginning to lose their greenness
Okra	Desirable size reached and the tips of which can be snapped readily
Upo, snake gourd, and dishrag gourd	Desirable size reached and thumbnail can still penetrate flesh readily (overmature if thumbnail cannot penetrate flesh readily)
Eggplant, bitter gourd, chayote or slicing cucumber	Desirable size reached but still tender (overmature if color dulls or changes and seeds are tough)

Sweet corn	Exudes milky sap when thumbnail penetrates kernel
Tomato	Seeds slipping when fruit is cut, or green color turning pink
Sweet pepper	Deep green color turning dull or red
Muskmelon	Easily separated from vine with a slight twist leaving clean cavity
Honeydew melon	Change in fruit color from a slight greenish white to cream; aroma noticeable
Watermelon	Color of lower part turning creamy yelow, dull hollow sound when thumped
Flower vegetables	
Cauliflower	Curd compact (overmature if flower cluster elongates and become loose)
Broccoli	Bud cluster compact (overmature if loose)
Leafy vegetables	
Lettuce	Big enough before flowering
Cabbage	Head compact (overmature if head cracks)
Celery	Big enough before it becomes pithy

Ripening follows or overlaps maturation, rendering the produce edible, as indicated by taste. Senescence is the last stage, characterised by natural degradation of the fruit or vegetable, as in loss of texture, flavour, etc. (senescence ends at the death of the tissue of the fruit). Some typical maturity indexes are described below:

Skin colour: This factor is commonly applied to fruits, since skin colour changes as fruit ripens or matures. Some fruits exhibit no perceptible colour change during maturation, depending on the type of fruit or vegetable. Assessment of harvest maturity by skin colour depends on the judgment of the harvester, but colour charts are available for cultivars, such as apples, tomatoes, peaches, chilli peppers, etc.

Optical methods: Light transmission properties can be used to measure the degree of maturity of fruits. These methods are based on the chlorophyll content of the fruit, which is reduced during maturation. The fruit is exposed to a bright light, which is then switched off so that the fruit is in total darkness. Next, a sensor measures the amount of light emitted from the fruit, which is proportional to its chlorophyll content and thus its maturity.

Shape: The shape of fruit can change during maturation and can be used as a characteristic to determine harvest maturity. For instance, a banana becomes more rounded in cross-sections and less angular as it develops on the plant. Mangoes also change shape during maturation.

As the mango matures on the tree the relationship between the shoulders of the fruit and the point at which the stalk is attached may change. The shoulders of immature mangoes slope away from the fruit stalk; however, on more mature mangoes the shoulders become level with the point of attachment, and with even more maturity the shoulders may be raised above this point.

Size: Changes in the size of a crop while growing are frequently used to determine the time of harvest. For example, partially mature cobs of Zea mays saccharata are marketed as sweet corn, while even less mature and thus smaller cobs are marketed as baby corn. For bananas, the width of individual fingers can be used to determine harvest maturity. Usually a finger is placed midway along the bunch and its maximum width is measured with callipers; this is referred to as the calliper grade.

Aroma: Most fruits synthesise volatile chemicals as they ripen. Such chemicals give fruit its characteristic odour and can be used to determine whether it is ripe or not. These doors may only be detectable by humans when a fruit is completely ripe, and therefore has limited use in commercial situations.

Fruit opening: Some fruits may develop toxic compounds during ripening, such as ackee tree fruit, which contains toxic levels of hypoglycine. The fruit splits when it is fully mature, revealing black seeds on yellow arils. At this stage, it has been shown to contain minimal amounts of hypoglycine or none at all. This creates a problem in marketing; because the fruit is so mature, it will have a very short post-harvest life.

Analysis of hypoglycine 'A' (hyp.) in ackee tree fruit revealed that the seed contained appreciable hyp. at all stages of maturity, at approximately 1000 ppm, while levels in the membrane mirrored those in the arils. This analysis supports earlier observations that unopened or partially opened ackee fruit should not be consumed, whereas fruit that opens naturally to over 15 mm of lobe separation poses little health hazard, provided the seed and membrane portions are removed. These observations agree with those of Brown et al. who stated that bright red, full sized ackee should never be forced open for human consumption.

Leaf changes: Leaf quality often determines when fruits and vegetables should be harvested. In root crops, the condition of the leaves can likewise indicate the condition of the crop below ground. For example, if potatoes are to be stored, then the optimum harvest time is soon after the leaves and stems have died. If harvested earlier, the skins will be less resistant to harvesting and handling damage and more prone to storage diseases.

Abscission: As part of the natural development of a fruit an abscission layer is formed in the pedicel. For example, in cantaloupe melons, harvesting before the abscission layer is fully developed results in inferior flavoured fruit, compared to those left on the vine for the full period.

Firmness: A fruit may change in texture during maturation, especially during ripening when it may become rapidly softer. Excessive loss of moisture may also affect the texture of crops. These textural changes are detected by touch, and the harvester may simply be able to gently squeeze the fruit and judge whether the crop can be harvested. Today sophisticated devices have been developed to measure texture in fruits and vegetables, for example, texture analysers and pressure testers; they are currently available for fruits and vegetables in various forms.

A force is applied to the surface of the fruit, allowing the probe of the penetrometer or texturometer to penetrate the fruit flesh, which then gives a reading on firmness. Hand held pressure testers could give variable results because the basis on which they are used to measure firmness is affected by the angle at which the force is applied. Two commonly used pressure testers to measure the firmness of fruits and vegetables are the Magness-Taylor and UC Fruit Firmness testers (Figure 1). A more elaborate test, but not necessarily more effective, uses instruments like the Instron Universal Testing Machine. It is necessary to specify the instrument and all settings used when reporting test pressure values or attempting to set standards.

Juice content: The juice content of many fruits increases as the fruit matures on the tree. To measure the juice content of a fruit, a representative sample of fruit is taken and then the juice extracted in a standard and specified manner.

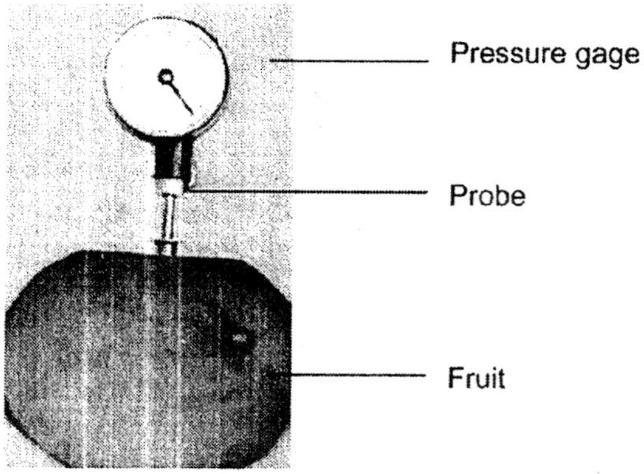

Figure 1. Pressure tester used to measure firmness of fruits and vegetables.

The juice volume is related to the original mass of juice, which is proportional to its maturity. The minimum values for citrus juices are presented in Table 1.

Table 1. Minimum juice values for mature citrus.

Citrus fruit	Minimum juice content (%)
Naval oranges	30
Other oranges	35
Grapefruit	35
Lemons	25
Mandarins	33
Clementines	40

Oil content and dry matter percentage: Oil content can be used to determine the maturity of fruits, such as avocados. According to the Agricultural Code in California, avocados at the time of harvest and at any time thereafter, shall not contain in weight less than 8%

oil per avocado, excluding skin and seed (Mexican or Guatemalan race cultivars). Thus, the oil content of an avocado is related to moisture content. The oil content is determined by weighing 5-10 g of avocado pulp and then extracting the oil with a solvent (e.g., benzene or petroleum ether) in a destillation column. This method has been successful for cultivars naturally high in oil content.

A round flask is used for the solvent. Heat is supplied with an electric plate and water recirculated to maintain a constant temperature during the extraction process (Figure 2). Extraction is performed using solvents such as petroleum ether, benzene, diethyl ether, etc., a process that takes between 4-6 h. After the extraction, the oil is recovered from the flask through evaporation of the water at 105°C in an oven until constant weight is achieved.

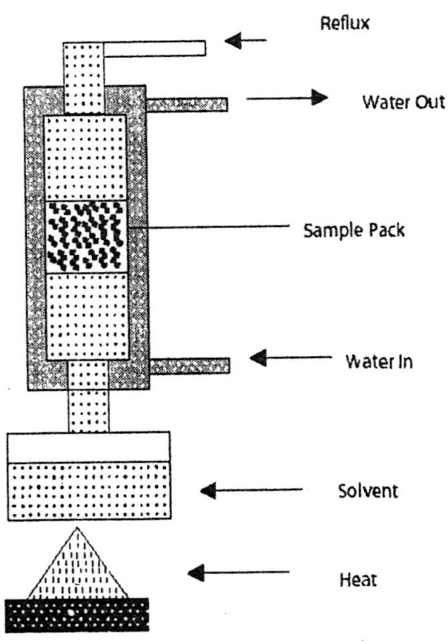

Figure 2. Distillation column used for oi' determination.

Moisture content: During the development of avocado fruit the oil content increases and moisture content rapidly decreases. The moisture levels required to obtain good acceptability of a variety of avocados cultivated in Chile are listed in Table 2.

Sugars: In climacteric fruits, carbohydrates accumulate during maturation in the form of starch. As the fruit ripens, starch is broken down into sugar. In non-climacteric fruits, sugar tends to accumulate during maturation. A quick method to measure the amount of sugar present in fruits is with a brix hydrometer or a refractometer. A drop of fruit juice is placed in the sample holder of the refractometer and a reading taken; this is equivalent to the total amount of soluble solids or sugar content. This factor is used in many parts of the world to specify maturity. The soluble solids content of fruit is also determined by shining light on the fruit or vegetable and measuring the amount transmitted. This is a laboratory technique however and might not be suitable for village level production.

Table 2. Moisture content of avocado fruit cultivated in Chile.

Cultivar	Moisture content (%)
Negra de la Cruz	80.1
Bacon	77.5
Zutano	80.5
Fuerte	77.9
Edranol	78.1
Hass	73.8
Gwen	78.4
Whitesell	79.1

Starch content: Measurement of starch content is a reliable technique used to determine maturity in pear cultivars.

The method involves cutting the fruit in two and dipping the cut pieces into a solution containing 4% potassium iodide and 1% iodine. The cut surfaces stain to a blue-black colour in places where starch is present. Starch converts into sugar as harvest time approaches. Harvest begins when the samples show that 65-70% of the cut surfaces have turned blue-black.

Acidity: In many fruits, the acidity changes during maturation and ripening, and in the case of citrus and other fruits, acidity reduces progressively as the fruit matures on the tree. Taking samples of such fruits, and extracting the juice and titrating it against a standard alkaline solution, gives a measure that can be related to optimum times of harvest. Normally, acidity is not taken as a measurement of fruit maturity by itself but in relation to soluble solids, giving what is termed the brix: acid ratio.

Specific gravity: Specific gravity is the relative gravity, or weight of solids or liquids, compared to pure distilled water at 62°F (16.7°C), which is considered unity. Specific gravity is obtained by comparing the weights of equal bulks of other bodies with the weight of water. In practice, the fruit or vegetable is weighed in air, then in pure water.

The weight in air divided by the weight in water gives the specific gravity. This will ensure a reliable measure of fruit maturity. As a fruit matures its specific gravity increases. This parameter is rarely used in practice to determine time of harvest, but could be used in cases where development of a suitable sampling technique is possible. It is used however to grade crops according to different maturities at post-harvest. This is done by placing the fruit in a tank of water, wherein those that float are less mature than those that sink.

Harvesting Practices

Harvesting practices should cause as little mechanical damage to produce as possible. Gentle digging, picking and handling will help reduce crop losses.

For some crops, a natural break point forms at the junction of the stem and the stalk when produce is mature. Harvesters should grasp the product firmly but gently and pull upward as illustrated below. Wearing cotton gloves, trimming fingernails, and removing jewelry such as rings and bracelets can help reduce mechanical damage during harvest.

If a small amount of leafy vegetables are being harvested for home use or for sale at a nearby roadside or farmers' market, a small tub of cold water can be useful for cooling the produce. The tub can be brought directly to the field and used by the picker as a field container. Clean water should be used with each lot of produce. Chilling leafy vegetables by using cold water at harvest will help maintain quality and prevent wilting.

Picking baskets, bags and buckets come in many sues and shapes. These harvesting containers can be made by sewing bags with openings on both ends, fitting fabric over the open bottom of ready-made baskets, fitting bags with adjustable harnesses, or by simply adding some carrying straps to a small basket.

Plastic crates are relatively expensive but are durable, reusable and easy to clean. When empty, they can be nested to save space in storage or transport.

Tools Used for Harvesting

Some fruits need to be clipped or cut from the parent plant. Clippers or knives should be kept well sharpened.

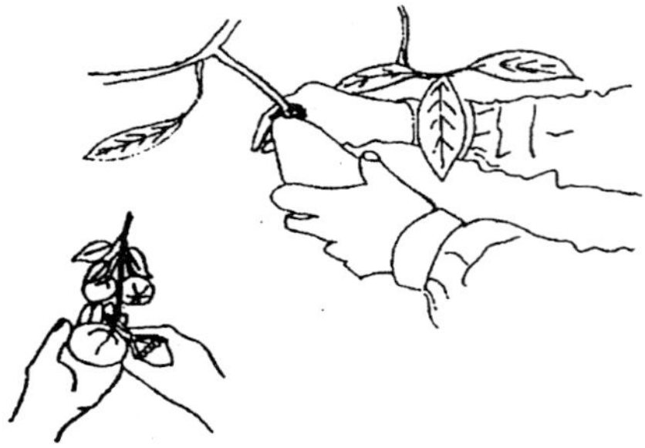

Penduncles, woody stems or spurs should be trimmed as close as possible to prevent fruit from damaging neighboring fruits during transport. Pruning shears are often used for harvesting fruits, some vegetables, and cut flowers. A variety of styles are available as hand held or pole models, including shears that cut and hold onto the stem of the cut product. This feature allows the picker to harvest without a catching bag and without dropping fruits.

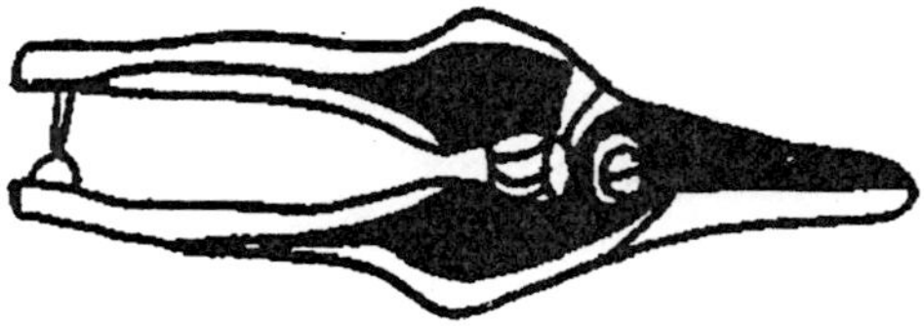

Straight bladed hand shears for fruits and flowers

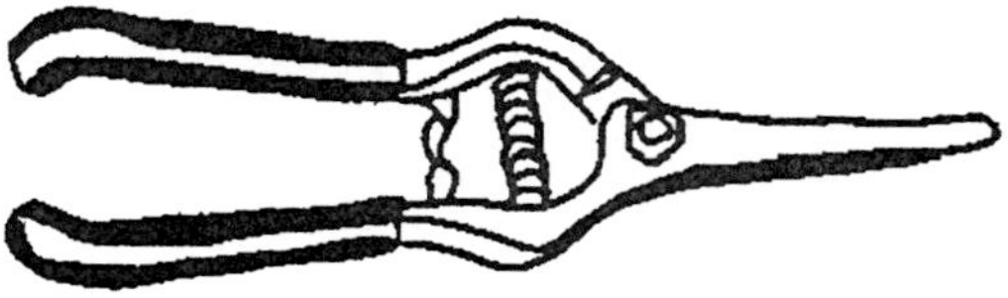

Thin curved blade for grapes and fruits

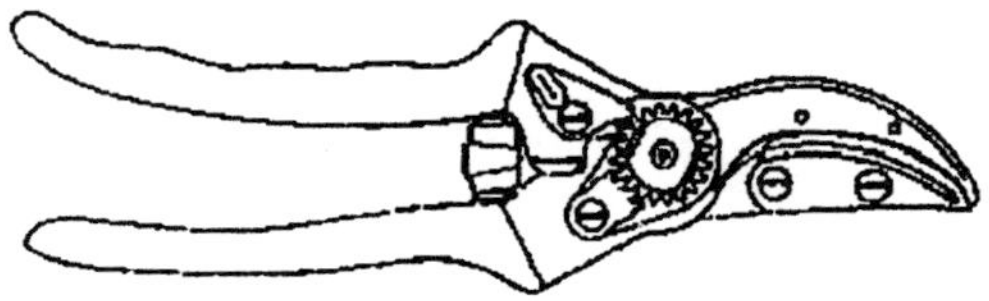

Cut and hold hand shears

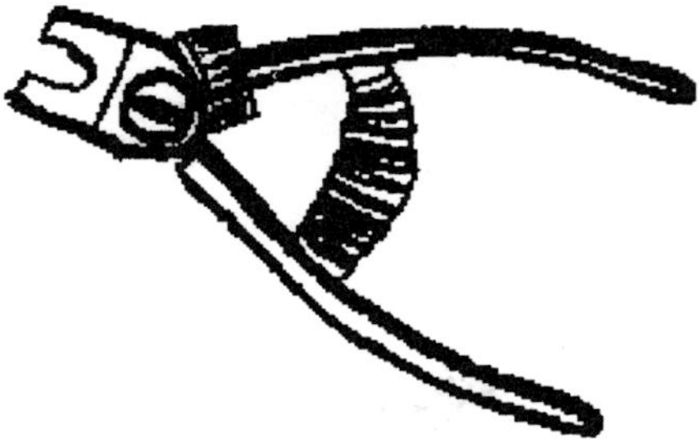

Clipper for citrus fruits

Pole mounted cut and hold picking shears:

Using a cutting tool attached to a long pole can aid picking of crops such as mangoes and avocados when the fruit is difficult to reach. Cutting edges should be kept sharpened and the catching bag should be relatively small The angle of the cutting edge and the shape of the catching bag can affect the quality of the fruit harvested, so it is important to check performance carefully before using any new tools.

Picking poles and catching sacks can be made by hand or purchased from horticultural supply companies.

The collection bags illustrated below were hand woven from strong cord or sewn from canvas. The hoop used as the basket rim and cutting edges can be fashioned from sheet metal, steel tubing or recycled scrap metal.

Fruit trees are sometimes quite tall and letting fruit fall to the ground when it is cut from the tree will cause severe bruising. If two pickers work together, one can clip or cut the fruit from the tree, and the other can use a sack to break its fall. The catcher supports the bag with his hands and one foot, catches the falling fruit, then lowers the far end of the bag to allow the fruit to roll safely to the ground.

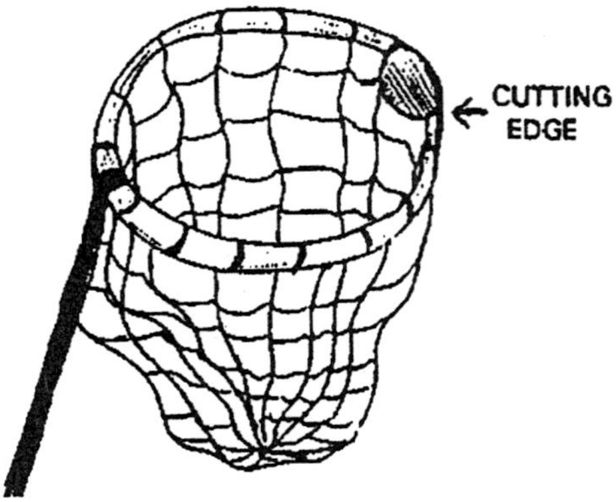

Hand woven collection bag

Unlike most nut crops, pistachio nuts should not be knocked to the ground during harvest because of their open shells and relatively high moisture content. The harvesting practice illustrated below can be used with pistachios and olives with good results. Plastic sheeting or canvas tarpaulins are spread below the tree being harvested, and trees are mechanically shaken or hand knocked (the branches hit with mauls) until the nuts drop. In the illustration below, two harvesters are gathering a sheet covered with produce.

Field Packing

When crops are field packed the picker harvests and then immediately packs the produce after minimal. handling Strawberries are generally field packed, since even a small amount of handling will damage these soft fruits. When lettuce is field packed, several wrapper leaves are left on the head to help cushion the produce during transport. A small cart can help reduce the amount of bending and lifting the picker has to do during harvest. The carts shown below have a single wheel in front, and can be pushed along the row ahead the e picker.

Field packing lettuce

A simple aid for field packers is a movable cart with a rack for boxes and a roof to provide shade. This cart is designed to be pushed along the outer edge of the small field where harvest is taking place.

This cart for field packing is designed to be pulled by a small tractor into the field when the crop is harvested. This type of cart can be used for field packing many types of crops. The roof folds down for easy transport, and opens up to provide a wide area of shade for the packers and the commodity. The cart design can be modified as needed to suit various products and different operations.

A self-propelled field pack system allows field workers to cut, trim, tie/wrap and pack in the field, thus eliminating the expense of operating a packing shed.

Transportation

When crops are harvested at some distance from the packinghouse, the produce must be transported before packing. The gravity driven conveyor system for bananas illustrated below provides an example of how handling can be minimised during preparation for market. Harvested bananas are carried to the platforms set up along the conveyor route, then lifted and hung from hooks attached to the wire. Transport speed is controlled by workers who lead the produce to the packinghouse at the bottom of the hill.

Berries picked for the fresh market (except blueberries and cranberries) are often mechanically harvested and usually packed into shipping containers. Careful harvesting, handling, and transporting of fruits and vegetables to packinghouses are necessary to preserve product quality.

Polyethylene bags: Clear polyethylene bags are used to pack banana bunches in the field, which are then transported to the packinghouse by means of mechanical cableways running through the banana plantation. This technique of packaging and transporting bananas reduces damage to the fruit caused by improper handling.

Plastic field boxes: These types of boxes are usually made of polyvinyl chloride, polypropylene, or polyethylene. They are durable and can last many years. Many are designed in such a way that they can nest inside each other when empty to facilitate transport, and can stack one on top of the other without crushing the fruit when full (Figure 3).

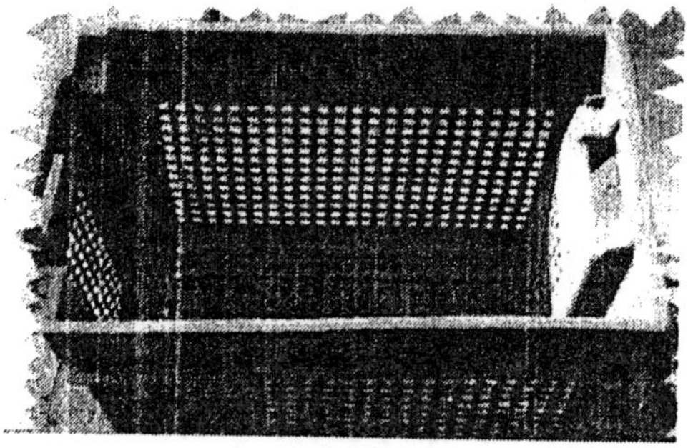

Figure 3. Plastic field boxes with nest/stack design.

Wooden field boxes: These boxes are made of thin pieces of wood bound together with wire. They come in two sizes: the bushel box with a volume of 2200 in^3 (36052 cm^3) and the half-bushel box. They are advantageous because they can be packed flat and are inexpensive, and thus could be non-returnable.

Figure 4. Typical wooden crate holding fresh tomatoes.

They have the disadvantage of providing little protection from mechanical damage to the produce during transport. Rigid wooden boxes of different capacities are commonly used to transport produce to the packinghouse or to market. (Figure 4).

Table 3. Approximate depth of bulk bins.

Commodity	Depth (cm)
Citrus	70
Pears, apples	50
Stone fruits	50
Tomatoes	40

Bulk bins: Bulk bins of 200-500 kg capacity are used for harvesting fresh fruits and vegetables. These bins are much more economical than the field boxes, both in terms of fruit carried per unit volume and durability, as well as in providing better protection to the product during transport to the packinghouse. They are made of wood and plastic materials. Dimensions for these bins in the

United States are 48 × 40 in, and 120 × 100 cm in metric system countries. Approximate depth of bulk bins depends on the type of fruit or vegetable being transported (Table 3).

Post-harvest Handling

In agriculture, post-harvest handling is the stage of crop production immediately following harvest, including cooling, cleaning, sorting and packing. The instant a crop is removed from the ground, or separated from its parent plant, it begins to deteriorate. Post-harvest treatment largely determines final quality, whether a crop is sold for fresh consumption, or used as an ingredient in a processed food product. The most important goals of post-harvest handling are keeping the product cool, to avoid moisture loss and slow down undesirable chemical changes, and avoiding physical damage such as bruising, to delay spoilage.

Sanitation is also an important factor, to reduce the possibility of pathogens that could be carried by fresh produce, for example, as residue from contaminated washing water. After the field, post-harvest processing is usually continued in a packing house. This can be a simple shed, providing shade and running water, or a large-scale, sophisticated, mechanised facility, with conveyor belts, automated sorting and packing stations, walk-in coolers and the like. In mechanised harvesting, processing may also begin as part of the actual harvest process, with initial cleaning and sorting performed by the harvesting machinery.

Initial post-harvest storage conditions are critical to maintaining quality. Each crop has an optimum range for storage temperature and humidity. Also, certain crops cannot be effectively stored together, as unwanted chemical interactions can result. Various methods of high-speed cooling, and sophisticated refrigerated and atmosphere-controlled environments, are employed to prolong freshness, particularly in large-scale operations.

Regardless of the scale of harvest, from home garden to industrialised farm, the basic principles of post-harvest handling for most crops are the same:

– handle with care to avoid damage (cutting, crushing, bruising)

– cool immediately and maintain in cool conditions

– cull (remove damaged items)

– cool immediately and maintain in cool conditions

– cull (remove damaged items)

Curing

When roots and tubers are to be stored for long periods, curing is necessary to extend the shelf life. The curing process involves the application of high temperatures and high relative humidity to the roots and tubers for long periods, in order to heal the skins wounded during harvesting. With this process a new protected layer of cells is formed. Initially the curing process is expensive, but in the long run, it is worthwhile.

Curing can be accomplished in the field or in curing structures conditioned for that purpose. Commodities such as yams can be cured in the field by piling them in a partially shaded area. Cut grass or straw can serve as insulating material while covering the pile with canvas, burlap, or woven grass matting. This covering will provide sufficient heat to reach high temperatures and high relative humidity. The stack can be left in this state for up to four days. Onions and garlic can be cured in the field in windrows or after being packed into large fibre or net sacks.

Modern curing systems have been implemented in housing conditioned with fans and heaters to produce the heat necessary for high temperatures and high relative humidity, as illustrated in the next page.

The fans are used to redistribute the heat to the lower part of the room where the produce is stored. Bulk bins are stacked with a gap of 10 to 15 cm between rows to allow adequate air passage. The system shown in Figure 5 can be used for curing onions; an exhaust opening near the ceiling must be provided for air recirculation. Care should be taken to prevent over-dryness of the onion bulbs.

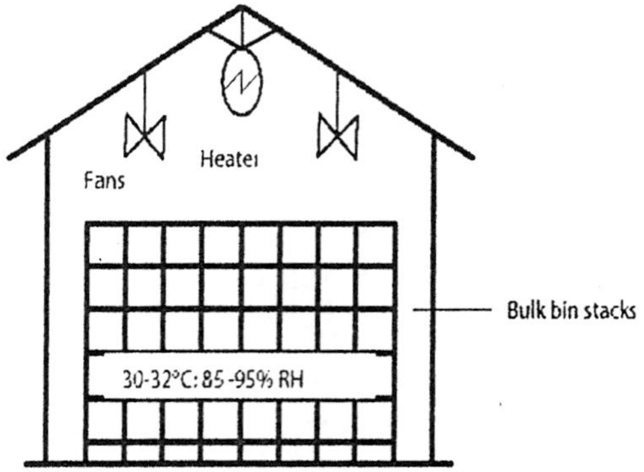

Figure 5. Typical curing houses for roots and tubers.

When extreme conditions in the field exist, such as heavy rain or flooded terrain, and curing facilities are not available, a temporary tent must be constructed from large tarpaulins or plastic sheets to cure the onions and avoid heavy loss. Heated air is forced into a hollow area at the centre of the produce-filled bins. Several fans are used to recirculate the warm air through the onions while curing.

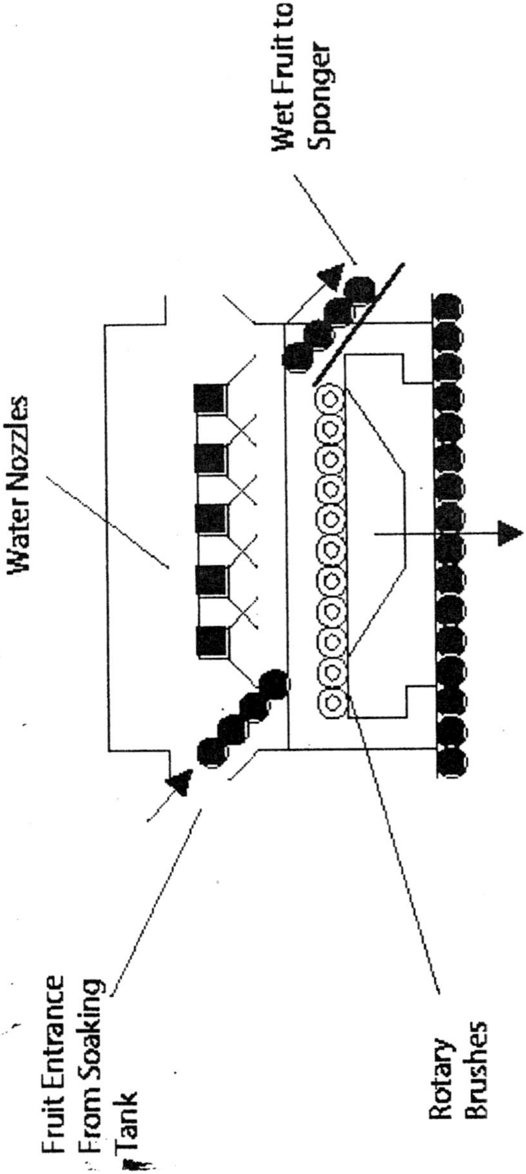

Figure 6. Typical produce washing machine.

Fruits and vegetables are subjected to preliminary treatments designed to improve appearance and maintain quality. These preparatory treatments include cleaning, disinfection, waxing, and adding of colour (some includes brand name stamping on individual fruits).

Cleaning: Most produce receives various chemical treatments such as spraying of insecticides and pesticides in the field. Most of these chemicals are poisonous to humans, even in small concentrations. Therefore, all traces of chemicals must be removed from produce before packing. As illustrated in Figure 6, the fruit or vegetable passes over rotary brushes where it is rotated and transported to the washing machine and exposed to the cleaning process from all sides.

From the washing machine, the fruit passes onto a set of rotary sponge rollers (similar to the rotary brushes). The rotary sponges remove most of the water on the fruit as it is rotated and transported through the sponger.

Disinfection: After washing fruits and vegetables, disinfectant agents are added to the soaking tank to avoid propagation of diseases among consecutive batches of produce. In a soaking tank, a typical solution for citrus fruit includes a mixture of various chemicals at specific concentration, pH, and temperature, as well as detergents and water softeners. Sodium-ortho-phenyl-phenate (SOPP) is an effective citrus disinfectant, but requires precise control of conditions in the tank.

Concentrations must be kept between 0.05 and 0.15%, with pH at 11.8 and temperature in the range of 43-48°C. Recommended soaking time is 3-5 minutes. Deviation from these recommendations may have disastrous effects on the produce, since the solution will be ineffective if the temperature or concentration is too low (Peleg, 1985).

Low concentrations of chlorine solution are also used as disinfectant for many vegetables. The advantage of this solution is that it does not leave a chemical residue on the product.

Artificial waxing: Artificial wax is applied to produce to replace the natural wax lost during washing of fruits or vegetables. This adds a bright sheen to the product. The function of artificial waxing of produce is summarised below:

- Provides a protective coating over entire surface.
- Seals small cracks and dents in the rind or skin.
- Seals off stem scars or base of petiole.
- Reduces moisture loss.
- Permits natural respiration.
- Extends shelf life.
- Enhances sales appeal.

Brand name application: Some distributors use ink or stickers to stamp a brand name or logo on each individual fruit. Ink is not permissible in some countries (e.g., Japan), but stickers are acceptable. Automatic machines for dispensing and applying pressure sensitive paper stickers are readily available. The advantage of stickers is that they can be easily peeled off.

Packaging

Modern packaging must comply with the following requirements:

a) The package must have sufficient mechanical strength to protect the contents during handling, transport, and stacking.

b) The packaging material must be free of chemical substances that could transfer to the produce and become toxic to man.

c) The package must meet handling and marketing requirements in terms of weight, size, and shape.

d) The package should allow rapid cooling of the contents. Furthermore, the permeability of plastic films to respiratory gases could also be important.

e) Mechanical strength of the package should be largely unaffected by moisture content (when wet) or high humidity conditions.

f) The security of the package or ease of opening and closing might be important in some marketing situations.

g) The package must either exclude light or be transparent.

h) The package should be appropriate for retail presentations.

i) The package should be designed for ease of disposal, re-use, or recycling.

j) Cost of the package in relation to value and the extent of contents protection required should be as low as possible.

Packages can be classified as follows:

- Flexible sacks; made of plastic jute, such as bags (small sacks) and nets (made of open mesh)
- Wooden crates
- Cartons (fibreboard boxes)
- Plastic crates

- Pallet boxes and shipping containers
- Baskets made of woven strips of leaves, bamboo, plastic, etc.

Nets are only suitable for hard produce such as coconuts and root crops (potatoes, onions, yams). Wooden crates are typically wire bound crates used for citrus fruits and potatoes, or wooden field crates used for softer produce like tomatoes. Wooden crates are resistant tc weather and more efficient for large fruits, such as watermelons and other melons, and generally have good ventilation. Disadvantages are that rough surfaces and splinters can cause damage to the produce, they can retain undesirable odours when painted, and raw wood can easily become contaminated with moulds.

Fibreboard boxes are used for tomato, cucumber, and ginger transport. They are easy to handle, light weight, come in different sizes, and come in a variety of colours that can make produce more attractive to consumers. They have some disadvantages, such as the effect of high humidity, which can weaken the box; neither are they waterproof, so wet products would need to be dried before packaging. These boxes are often of lower strength compared to wooden or plastic crates, although multiple thickness trays are very widely used. They can come flat packed with ventilation holes and grab handles, making a cheap attractive alternative that is very popular.

Care should be taken that holes on the surface (top and sides) of the box allow adequate ventilation for the produce and prevent heat generation, which can cause rapid product deterioration. Plastic crates are expensive but last longer than wooden or carton crates. They are easy to clean due to their smooth surface and are hard in strength, giving protection to products.

Plastic crates (Figure 7) can be used many times, reducing the cost of transport. They are available in different sizes and colours and are resistant to adverse weather conditions. However, plastic crates can damage some soft produce due to their hard surfaces, thus liners are recommended when using such crates.

Pallet boxes are very efficient for transporting produce from the field to the packinghouse or for handling produce in the packinghouse. Pallet boxes have a standard floor size (1200 × 1000 mm) and depending on the commodity have standard heights. Advantages of the pallet box are that it reduces the labour and cost of loading, filling, and unloading; reduces space for storage; and increases speed of mechanical harvest.

Figure 7. Typical plastic crate holding fresh oranges.

The major disadvantage is that the return volume of most pallet boxes is the same as the full load. Higher investment is also required for the forklift truck, trailer, and handling systems to empty the boxes. They are not affordable to small producers because of high, initial capital investment.

Cooling Methods

Several methods of cooling are applied to produce after harvesting to extend shelf life and maintain a fresh-like quality. Some of the low temperature treatments are unsuitable for simple rural or village treatment but are included for consideration as follows:

Precooling: Fruit is precooled when its temperature is reduced from 3 to 6°C (5 to 10°F) and is cool enough for safe transport. Precooling may be done with cold air, cold water (hydrocooling), direct contact with ice, or by evaporation of water from the product under a partial vacuum (vacuum cooling). A combination of cooled air and water in the form of a mist called hyraircooling is an innovation in cooling of vegetables.

Air precooling: Precooling of fruits with cold air is the most common practice. It can be done in refrigerator cars, storage rooms, tunnels, or forced air-coolers (air is forced to pass through the container via baffles and pressure differences).

Icing: Ice is commonly added to boxes of produce by placing a layer of crushed ice directly on the top of the crop. An ice slurry can be applied in the following proportion: 60% finely crushed ice, 40% water, and 0.1% sodium chloride to lower the melting point. The water to ice ratio may vary from 1:1 to 1:4.

Room cooling: This method involves placing the crop in cold storage. The type of room used may vary, but generally consists of a refrigeration unit in which cold air is passed through a fan. The circulation may be such that air is blown across the top of the room and falls through the crop by convection. The main advantage is cost because no specific facility is required.

Forced air-cooling: The principle behind this type of precooling is to place the crop into a room where cold air is directed through the crop after flowing over various refrigerated metal coils or pipes. Forced air-cooling systems blow air at a high velocity leading to desiccation of the crop. To minimise this effect, various methods of humidifying the cooling air have been designed such as blowing the air through cold water sprays.

Hydrocooling: The transmission of heat from a solid to a liquid is faster than the transmission of heat from a solid to a gas. Therefore, cooling of crops with cooled water can occur quickly and results in zero loss of weight. To achieve high performance, the crop is submerged in cold water, which is constantly circulated through a heat exchanger. When crops are transported around the packhouse in water, the transport can incorporate a hydrocooler. This system has the advantage wherein the speed of the conveyer can be adjusted to the time required to cool the produce.

Hydrocooling has a further advantage over other precooling methods in that it can help clean the produce. Chlorinated water can be used to avoid spoilage of the crop. Hydrocooling is commonly used for vegetables, such as asparagus, celery, sweet corn, radishes, and carrots, but it is seldom used for fruits.

Vacuum cooling: Cooling in this case is achieved with the latent heat of vaporisation rather than conduction. At normal air pressure (760·mmHg) water will boil at 100°C. As air pressure is reduced so is the boiling point of water, and at 4.6 mmHg water boils at 0°C. For every 5 or 6°C reduction in temperature, under these conditions, the crop loses about 1% of its weight. This weight loss may be minimised by spraying the produce with water either before enclosing it in the vacuum chamber or towards the end of the vacuum cooling operation (hydrovacuum cooling).

The speed and effectiveness of cooling is related to the ratio between the mass of the crop and its surface area. This method is particularly suitable for leaf crops such as lettuce. Crops like tomatoes having a relatively thick wax cuticle are not suitable for vacuum cooling.

Storage Temperature

There is no ideal storage for all fruits and vegetables, because their response to reduced temperatures varies widely. The importance of factors such as mould growth and chilling injuries must be taken into account, as well as the required length of storage. Storage temperature for fruits and vegetables can range from -1 to 13°C, depending on their perishability. Extremely perishable fruits such as apricots, berries, cherries, figs, watermelons can be stored at -1 to 4°C for 1-5 weeks; less perishable fruits such as mandarin, nectarine, ripe or green pineapple can be stored at 5-9°C for 2-5 weeks; bananas at 10°C for 1-2 weeks and green bananas at 13°C for 1-2 weeks.

Highly perishable vegetables can be stored up to 4 weeks such as asparagus, beans, broccoli, and Brussels sprouts at -1-4°C for 1-4 weeks; cauliflower at 5-9°C for 2-4 weeks. Green tomato is less perishable and can be stored at 10°C for 3-6 weeks and non-perishable vegetables such as carrots, onions, potatoes and parsnips can be stored at 5-9°C for 12-28 weeks. Similarly, sweet potatoes can be stored at 10°C for 16-24 weeks.

The storage life of produce is highly variable and related to the respiration rate; there is an inverse relation between respiration rate and storage life in that produce with low respiration generally keeps longer. For example, the respiration rate of a very perishable fruit like ripe banana is 200 mL CO2.kg-1h-1 at 15°C, compared to a non-perishable fruit such as apple, which has a respiration rate of 25 mL CO2.kg-1 h-1 at 15°C.

Exposure of fruits and vegetables to high temperatures during post-harvest reduces their storage or marketable life. This is because as living material, their metabolic rate is normally higher with higher temperatures. High temperature treatments are beneficial in curing root crops, drying bulb crops, and controlling diseases and pests in some fruits. Many fruits are exposed to high temperatures in combination with ethylene (or another suitable gas) to initiate or improve ripening or skin colour.

Storage

The marketable life of most fresh vegetables can be extended by prompt storage in an environment that maintains product quality. The desired environment can be obtained in facilities where temperature, air circulation, relative humidity, and sometimes atmosphere composition can be controlled. Storage rooms can be grouped accordingly as those requiring refrigeration and those that do not. Storage rooms and methods not requiring refrigeration include: in situ, sand, coir, pits, clamps, windbreaks, cellars, barns, evaporative cooling, and night ventilation:

In situ: This method of storing fruits and vegetables involves delaying the harvest until the crop is required. It can be used in some cases with root crops, such as cassava, but means that the land on which the crop was grown will remain occupied and a new crop cannot be planted. In colder climates, the crop may be exposed to freezing and chilling injury.

Sand or coir: This storage technique is used in countries like India to store potatoes for longer periods of time, which involves covering the commodity under ground with sand. Pits or trenches are dug at the edges of the field where the crop has been grown. Usually pits are placed at the highest point in the field, especially in regions of high rainfall. The pit or trench is lined with straw or other organic material and filled with the crop being stored, then covered with a layer of organic material followed by a layer of soil. Holes are created with straw at the top to allow for air ventilation, as lack of ventilation may cause problems with rotting of the crop.

Clamps: This has been a traditional method for storing potatoes in some parts of the world, such as Great Britain. A common design uses an area of land at the side of the field. The width of the clamp is about 1 to 2.5 m. The dimensions are marked out and the potatoes piled on the ground in an elongated conical heap. Sometimes straw is laid on the soil before the potatoes.

The central height of the heap depends on its angle of repose, which is about one third the width of the clump. At the top, straw is bent over the ridge so that rain will tend to run off the structure. Straw thickness should be from 15-25 cm when compressed. After two weeks, the clamp is covered with soil to a depth of 15-20 cm, but this may vary depending on the climate.

Windbreaks are constructed by driving wooden stakes into the ground in two parallel rows about 1 m apart. A wooden platform is built between the stakes about 30 cm from the ground, often made from wooden boxes. Chicken wire is affixed between the stakes and across both ends of the windbreak. This method is used in Britain to store onions.

Cellars: These underground or partly underground rooms are often beneath a house. This location has good insulation, providing cooling in warm ambient conditions and protection from excessively low temperatures in cold climates. Cellars have traditionally been used at domestic scale in Britain to store apples, cabbages, onions, and potatoes during winter.

Barns: A barn is a farm building for sheltering, processing, and storing agricultural products, animals, and implements. Although there is no precise scale or measure for the type or size of the building, the term barn is usually reserved for the largest or most important structure on any particular farm. Smaller or minor agricultural buildings are often labelled sheds or outbuildings and are normally used to house smaller implements or activities.

Evaporative cooling: When water evaporates from the liquid phase into the vapour phase energy is required. This principle can be used to cool stores by first passing the air introduced into the storage room through a pad of water. The degree of cooling depends on the original humidity of the air and the efficiency of the evaporating surface. If the ambient air has low humidity and is humidified to around 100% RH, then a large reduction in temperature will be achieved. This can provide cool moist conditions during storage.

Night ventilation: In hot climates, the variation between day and night temperatures can be used to keep stores cool. The storage room should be well insulated when the crop is placed inside. A fan is built into the store room, which is switched on when the outside temperature at night becomes lower than the temperature within. The fan switches off when the temperatures equalise. The fan is controlled by a differential thermostat, which constantly compares the outside air temperature with the internal storage temperature. This method is used to store bulk onions.

Controlled atmospheres are made of gastight chambers with insulated walls, ceiling, and floor. They are increasingly common for fruit storage at larger scale. Depending on the species and variety, various blends of O2, CO2, and N2 are required. Low content O2 atmospheres (0.8 to 1.5%), called ULO (Ultra -Low Oxygen) atmospheres, are used for fruits with long storage lives (e.g., apples).

Pest Control

Crops may be immersed in hot water before storage or marketing to control disease. A common disease of fruits known as anthracnose, caused by the infection of fungus Colletotrychum spp. can be successfully controlled in this way. Combining appropriate doses of fungicides with hot water is often effective in controlling disease in fruits after harvesting.

Fruit and vegetable decay is also caused by storage conditions. Too low temperatures can cause injury during refrigeration of fruits and vegetables. High temperatures can cause softening of tissues and promote bacterial diseases. The damage that microorganisms inflict on fresh fruits and vegetables is mainly in the physical loss of edible matter, which may be partial or total.

2

Microbial Hazards in Fruits and Vegetables

Minimal processing of fruits and vegetables presents unique challenges, because cutting and slicing remove the natural protective barriers of the intact plant. Thus, implementing Hazard Analysis and Critical Control Points (HACCP) programs in high care facilities adds more assurance of food safety.

Scientists have used surrogate organisms to imitate the survival of microbial pathogens in fresh produce and these studies could provide significant insight into controlling the spread of pathogens in the industry. Produce is moved globally to supply year-round demands and improvement in traceability methods would help in epidemiological investigations. Outbreak data have been limited to just a few industrialised nations because only these countries have active surveillance systems for monitoring. Thus, the generation of more epidemiological data on produce-related foodborne illness worldwide will help determine true levels of illness.

Biological hazards are of great concern to the fresh produce industry. They can be classified into spore-forming bacteria, non-spore-forming bacteria, viruses, and parasites. Certain bacteria form spores to withstand environmental stress conditions such as high heat on freezing. Spore-forming organisms can attach to vegetables grown near the soil. Examples of these organisms include Bacillus cereus, Clostridium perfringens, and Clostridium botulinum. Maintaining refrigeration temperatures at less than 58C and promoting oxygen in packaging would reduce the risk of vegetative cell formation and the production of dangerous toxins that cause illness.

Non-spore-forming bacteria such as enterotoxigenic and enterohemorrhagic Escherichia coli, Campylobacter jejuni, Listeria monocytogenes, Salmonella, Shigella spp., Staphylococcus aureus, and Vibrio spp. could contaminate fresh produce by cross contact with humans or animals carrying these organisms. The fecal-oral route is possibly the main mechanism of transfer. All of these bacteria have been associated with publicised fresh produce foodborne outbreaks of public health significance. The transfer of these organisms could be controlled by practising good personal hygiene, cleaning food contact surfaces, and always using potable water when water is required.

Foodborne viruses require a living host in which to grow and reproduce. Viruses tend to move from one food to another, from water supply to food, or from food handler to food. Hepatitis A, Norwalk virus, and rotavirus are viruses of public health significance. Hepatitis A has been isolated in vegetables washed with nonpotable water. A food worker can carry the organism virus for up to six weeks and contaminate food and other workers without any knowledge of signs and symptoms. The Norwalk virus and rotavirus have been associated with many foodborne infections.

Raw fruits and vegetables washed with contaminated water were implicated in some outbreaks. These viruses are transmitted by person-to-person contact and by fecal contamination. Practising good personal hygiene and controlling staff carrying the virus are measures that could possibly eliminate foodborne illness. Parasitic protozoa include Cyclospora cayetanensis, Giardia lamblia, and Cryptosporidium parvum. They are single-cell microorganisms that must live on or inside a host to survive. These parasites may be transmitted via contaminated water, by person-to-person contact, and by fecal contamination.

Use of potable water for operations is critical. In the last 15 years, knowledge of foodborne disease epidemiology evolved while the fresh fruit and vegetable industry was undergoing notable changes. Factors increasing the risk of foodborne illness associated with fresh produce include the following:

- Modifications in agronomic practices, processing and packaging technologies;
- Global marketing strategies allowing fresh produce supply to consumers with a wide variety of products, year round;
- Changes in population demographics; and
- Changes in food consumption patterns.

Increased awareness because of unique epidemiologic surveillance programs and increased media attention has contributed to better documentation of foodborne illness. Numerous pathogens have been isolated from a wide variety of fresh fruits and vegetables. It is important to note that the number of samples in each study varied substantially. Although not all of the pathogens have been associated with produce-related foodborne disease outbreaks, they are all capable of causing illness. In the United States, a specific etiologic agent was identified for 187 produceassociated outbreaks during the years 1990-2002. Among these outbreaks, 102 (55%) were caused by bacteria, 68 (36%) were caused by viruses, and 17 (1%) were caused by parasites.

Among the bacterial agents, Salmonella accounted for 60% of outbreaks, and pathogenic E. coli was responsible for 25% of bacterial outbreaks. Norovirus caused a majority of viral outbreaks, accounting for over 80% of cases. The apparent prevalence of norovirus has increased possibly as a result of improved surveillance and detection methods. Cyclospora caused the majority (65%) of parasitic produce-associated outbreaks. Over 40% of the outbreaks were caused by salad items (including lettuce and tomatoes), whereas fruit and fruit salads comprised 13% of the outbreaks. Melons, including cantaloupe, honeydew, and watermelon, represented 12% of produce-associated outbreaks, and sprouts comprised 10% of the outbreaks.

Hazards in Production

The ground where product is grown plays a vital role in safety of the product. If the area has a history of use for chemical waste or the processing of biosolids, this would present a potential source of contamination for crops. It is important to know the land history and the time required for the area to lay fodder, thus reducing the level of contaminants in the soil. Adjacent land use also affects the safety of the crop grown. If fruit and vegetables are grown next to an animal-rearing operation, there is a potential for product to become contaminated by animals.

These animals may physically enter fields. Waste, high winds, and run-off from the animal operation may contaminate crops. The decision to grow next to a potentially hazardous location should be followed up with a risk assessment and the implementation of preventive measures to control risks identified. Sloping land from an adjacent field could be curbed by digging a ditch along the full length of the field to catch run-off water. Physical barriers and trenches may also prevent unwanted animal entry into fresh produce operations.

The importance of adjacent land was demonstrated in the first documented outbreak of Escherichia coli O157:H7 infection associated with a treated municipal water supply in Canada in May-June 2000. This was the largest multibacterial waterborne outbreak in Canada at that time. The Walkerton residents who became ill numbered approximately 1286. Researchers confirmed that a well was subject to surface water contamination and elevated turbidity.

Environmental testing of 13 livestock farms within a 4 km radius of the three wells identified human bacterial pathogens in animal manure on all but two farms. On nine farms, Campylobacter spp. were identified, on two farms both E. coli O157:H7 and Campylobacter spp. were found; this included a farm adjacent to the affected well. The evidence suggested that the pathogens that entered the well were likely to have originated from cattle manure on this farm.

Fruit and vegetable growers and packers are discouraged from keeping animals because they represent a source of product contamination. Domestic animals such as chickens, dogs, and horses can contaminate crop with fecal droppings if they pass through growing areas. Nonfarm animals such as deer, other mammals, and birds can serve as reservoirs for pathogens. Wild animals present a unique hazard to fruit and vegetable growers and these animals are difficult to control. Droppings from birds, deer, and other wild animals present a risk. It is not economical in some instances to fence large production areas, but small farms may be fenced to keep out wild animals.

Other physical barriers such as mounds, diversion berms, vegetative buffers, and ditches, provide protection from animals. Growers use scarecrows, reflective strips, and gunshots to ward off birds and pests from crop. Mechanical traps have been used to catch fieldmice in lettuce-growing operations in Yuma, Arizona. Growers implemented this emergency measure to alleviate a problem of mice in lettuce bound for the prepared salad industry. This measure was simple, yet effective in preserving product integrity.

An increased demand for organically grown produce promotes the use of alternative measures to protect plants from pests, mites, and fungi. In addition, organic fertilisers such as animal manure could introduce fecal pathogens to fresh produce if manure is not aged and treated before application. Treated manure or biosolids can be beneficial soil amendments. Growers have to manage these amendments very well to minimise contamination of fruits and vegetable with pathogenic fecal microorganisms.

Composting can effectively reduce pathogens and parasites commonly found in manure as well as those that have mutated into different strains with new abilities, like surviving in acidic environments. Once certain time and temperature criteria are achieved, E. coli and Salmonellae in the compost are nearly eliminated (99.9999 percent kill rate). The pathogen-reduction criteria include a temperature of at least 1318F for three consecutive days in an aerated pile or 1318F for two weeks in the hot zones of a windrow pile with five turnings. This process can kill nearly all pathogenic microbes and still maintain populations of beneficial ones.

Applying raw manure to a vegetable-growing operation could cause product contamination. In addition, manure piles stored next to growing operations may also be a problem because of run off. Organic growers use a great deal of treated manure as fertiliser and soil amendment. Organic growers must be vigilant not to use fresh manure, because this would increase the potential for product contamination. Manure application should be carried out well in advance of harvest so there is sufficient time for potential pathogens to break down naturally.

Water may be used throughout the growing and harvesting of fresh fruit and vegetables. Irrigation water and water used for application of plant protection product, fertiliser, and frost protection should not introduce risk of pathogens to product. The source of agricultural water could determine the final safety of the food product. Municipal water or wells are generally safe. In the United States, municipal water is usually treated and tested routinely to monitor microbial contamination. Wells with sound casing and absence of leaks should provide water free of pathogens.

Many growers also draw water from open water systems in areas where water is scarce. If effluent water from sewage plants is used in hydroponics plant production, the quality of this water is a concern for introduction of pathogens in edible food. The microflora of these systems changes daily and it is unknown what quality of water exists from one day to the next. Irrigation practices dictate product safety when using open systems like rivers and canals. Drip irrigation or furrow irrigation would bring water to the root of the plant without making contact with leaves or other edible portions.

In the early growing stages, it is possible to use sprinklers to wet young plants. The time interval between planting and harvesting is sufficiently long to expect pathogens present to die off. Greenhouse production is generally well protected from the environment, but is seen to favour the survival of human pathogens. Special issues of control of pathogens in greenhouses concern the fact that all nutrients are delivered to plants in liquid form. If water used for mixing agricultural chemicals becomes contaminated in any way, the implications for the greenhouse are extensive. The pathogen would spread over a wide range of product on which it was applied. The conditions in the greenhouse—warm temperature (.218C), moisture (.70 percent relative humidity), and light—would encourage the growth of pathogens.

Contamination in Harvesting

Harvesting is a critical period in the fresh produce chain, with potential for most product contamination. Employees in farming operations play a key role in maintaining the integrity of product grown. Their safe and hygienic practices—from land preparation, planting, weeding, and pruning, to harvesting—could influence whether product becomes contaminated. Personal hygiene, including hand washing all along the food chain, is critical in reducing or eliminating contamination with fecal pathogens.

Ill employees harvesting by hand could contaminate product bound for the consumer if care is not taken in basic hygiene. Growers may use portable toilet facilities that follow workers in large fields to provide sanitary facilities. As part of Good Agricultural Practices (GAPs), companies train field employees in the basics of using toilet facilities instead of the field, where product may be contaminated. Employees are trained in proper hand-washing techniques. In the United States, the Occupational Health and Safety Act (29) CFR 1928.110, subpart I, regulates the use of portable toilets.

The Act specifies that the following practices should be adopted:

- One toilet should be provided for every 20 workers.
- The field sanitation unit should be placed within one quarter mile of the harvest crew.
- Sanitary tissue, and soap and potable water for hand washing should be provided.
- A means of hand drying, preferably disposable paper towels, should be provided.
- Sanitary facilities in the field must be managed effectively so they do not become a source of contaminated product. A professional company should be used to empty and clean sanitary facilities.
- Growers should have a containment plan in place in the unlikely event of a spillage.

Wounds are hosts to pathogenic bacteria and direct food contact should be avoided. Workers who receive minor cuts while harvesting should only continue to work if the cut is well protected using a plaster/bandage, and further covered with gloves. If gloves are used, they should not cause product contamination. Thus, gloves should be changed regularly or when damaged. Equipment used for harvesting may also be a source of product contamination. Utensils should be cleaned and sanitised regularly to avoid cross-contamination.

Some growers now use a sanitising dip in the field during harvest so knives and other tools can be cleaned during the break and before start of harvest. Field containers should be washed routinely to avoid a build-up of debris. Some packhouses are equipped with crate washers that comprise a tunnel full of powerful water jets and in some cases an initial spray of foam cleaning detergent.

Because produce crates are taken to the field daily, it is important not to create any instance of crosscontamination of produce by using dirty packaging. Minimal processing of produce items like romaine hearts and head lettuce may be carried out in the field. Product is harvested, rinsed in the field on harvest rigs, and then packed in consumer packages and boxes. The harvest rigs should be cleaned and sanitised daily, because these products are advertised as "ready-to-eat". Any wash water used on harvest rigs should be potable and suitable for its application.

Utensils and vehicles used to transport packaged fruit and vegetables should be dedicated for this purpose, because cross-contamination could occur if other products are transported in the same vessels. A precautionary measure would be to wash and sanitise containers before use.

Post-harvest Handling

Packhouses. After product is harvested, it should be protected to prevent any cross-contamination. Packing facilities should be cleaned and well maintained to reduce the introduction of harmful microorganisms to product. Some growers move product from the field in large bins, which are taken to the packhouse for selection, grading, and repacking. No matter what method of packing is used, care must be taken with product. Cartons and other empty containers should not be stored uncovered in fields, because dust and animals could contaminate product.

Rigs and utensils used for packing should be cleaned and monitored daily. Packhouses, whether open or enclosed, should be cleaned and protected to deter pest entry and possible product contamination. Harvest storage facilities, containers, or bins should be cleaned regularly, based on a set program. Poor sanitation in the packhouse could lead to the formation of biofilms.

Biofilms are layers of bacteria that attach to surfaces like stainless steel and plastic, and also attach to each other with the help of polymeric materials. The biofilms trap other bacteria, debris and nutrients. Poor sanitation programs cause biofilms to build up and become established. Nonpathogenic and pathogenic bacteria can form biofilms. Organisms in the film tend to be resistant to cleaners and sanitisers, as well as heat treatment. Thus, in the food industry a sound sanitation program is needed so that biofilms do not become entrenched on food contact surfaces.

Good sanitation practices enhance a company's food-safety program. An important step is to provide training in sanitation to a wide base of employees, even those outside the sanitation department. In daily sanitation programs, seven steps could be followed to ensure clean equipment:

- Dry cleaning to remove gross debris from equipment and floors;
- Prerinsing to remove debris from surfaces;

– Using soap and scrubbing equipment on surfaces and floors;
– Postrinsing to remove soap;
– Removing standing water and reassembling equipment if necessary;
– Inspecting cleaned area and recleaning with detergent if necessary;
– Sanitising equipment and floors in high-care facilities.

Many companies use chlorine-based cleaning chemicals. For effective cleaning, some chemicals could require a specific residence time on equipment. Sanitisers include halogen-based compounds and quaternary ammonium-based compounds. For a review on food plant cleaning and sanitising, the reader is referred to Redemann. Pestmanagement could prevent product contamination, recall, and other loss of productivity. Acompany should have a preventive program of pest control so that problems never develop.

The packhouse manager should assess the risk in the plant to determine the level of prevention needed. Completely closed packhouses may have sophisticated pest control programs using a certified contract company. Windows and air vents should be screened and facilities kept free of debris. Any pest-control program implemented should be monitored and documented regularly to protect product.

Water Flumes. Where water is used to transport produce from one part of the packhouse to the next, the quality of that water will determine the quality and safety of the product being packed. Primary water for rinsing to remove dirt may be of agricultural grade. However, potable water should be used for subsequent rinsing steps. Fresh-cut rinsing operations may use automated chlorination and acidification systems to control and monitor water quality.

Flume channels used for sorting and grading product should be cleaned according to a planned sanitation program, thus preventing the build-up of debris from recycled water and avoiding product contamination. Where water is recycled, it should be treated to reduce the build up of microorganisms. Many companies use chlorine or other disinfecting chemicals to control the microbial load. Packers and processors use chlorine dioxide, chlorine (hypochlorous acid), UV light, ozone, hydrogen peroxide, peroxyacetic acid, and other sanitisers.

The standard sanitiser used in packhouses in the United States is chlorine. This could be in the form of sodium hypochlorite granules, tablets, or liquid. Water has been demonstrated to enter tomatoes through the stem end in a water dunk tank. This may be caused by the temperature differential of the produce item harvested in the field and the cooler water in the packhouse. At intake, any microorganisms in the water would also enter the produce item. Thus, it is imperative that any wash water in the plant be potable water.

Any pathogens in water could become the source of a foodborne outbreak. A multi-State outbreak of Salmonella enterica Serotype Newport infection was linked to mango consumption in December 1999. Traceback of the implicated mangoes led to a single Brazilian farm, where hot water treatment was identified as a possible point of contamination. Hot water treatment was a new process introduced to prevent importation of an agricultural pest, the Mediterranean fruit fly. Contaminated water caused product contamination. This outbreak highlighted the potential global health impact of foodborne diseases and newly implemented food processes and the vulnerability of product placed in contaminated water.

Employees. Packhouse employees should be trained in safe food-handling practices. Product safety of final consumer packs is directly influenced by the handling practices in the packhouse. Employees should receive training on the proper use of toilets, hand-washing procedures, use of protective clothing, and headgear to avoid product contamination. Many foodborne outbreaks have been linked to a sick employee who transferred pathogens via the fecal-oral route. Employees shedding pathogens in diarrhea may not wash their hands adequately before handling food. This is one possible route of contamination of fresh produce.

Storage and Distribution

Refrigeration temperatures in storage and distribution are crucial to maintaining product quality. These temperatures also reduce the proliferation of human pathogens if they are present on produce items. Refrigeration units are thought to spread mold throughout warehouses, and routine servicing of air filters and refrigeration systems is required. As cold air systems blow mold spores into the air, there is also the risk that pathogens may be spread along with the spores from one pallet to the other.

Pest control programs are necessary at any storage facility. A basic rodent control program would reduce the presence of pests that harbor harmful microorganisms forming a potential hazard to food. A contract company or trained individual should monitor traps or bait stations regularly and document any pest activity observed.

Vehicles and containers used to transport fresh produce could also be sources of potential contamination.

Vehicles used to transport fresh produce should be clean and free of odors, dirt, and debris before loading. The ideal situation would be to use dedicated containers and vehicles for each application; that is, produce, meat, or refuse should be transported by separate means. The produce containers and vehicles should also be cleaned routinely as part of Good Hygienic Practices (GHPs) to prevent contamination between loads.

Good hygienic and cleaning practices ensure product safety when loading or during inspections. The temperature of transport would also determine the potential for growth of pathogens. Thus, refrigeration temperatures are used to transport many produce items. The cold temperature helps to preserve product quality as well as safety.

Fresh-Cut Fruit and Vegetables

The fresh-cut fruit and vegetable industry in the United States grew from supplying quick-serve restaurant chains to providing consumer-size convenience products. From the 1980s to the 2000s, bagged salads and other convenience items experienced tremendous growth in demand. Sales of bagged salads topped $3 billion in 2004, while whole peeled carrots reached about $1 billion. Fresh-cut processors have worked with growers, retailers, and food service operations to ensure raw material and finished product are handled safely.

When fresh produce is washed, cut, and sliced, the natural defence mechanisms on the plant material are removed. The high level of handling increases the potential for product to be contaminated by microorganisms in the environment. Because there is no kill step in fresh-cut processing, using HACCP is imperative in maintaining food safety standards with preventive programs. Growers implement Good Agricultural Practices (GAPs) reducing risks during growing and harvesting.

Retailers and food service operations implement the principles of HACCP to manage food safety in their operations. In the United States, the International Fresh Cut Produce Association (IFPA) published Food Safety Guidelines for fresh-cut food processors. Guidance documents produced include a model HACCP plan, best practice guidelines for activities, a model food allergen plan, and a Sanitary Equipment Buying Guide and Development Checklist. The IFPA has also developed a GAP program in conjunction with the Western Growers' Association. The industry is thus equipped with proactive guidance to produce safe food.

The primary cause of foodborne illness in the United States is thought to be the mishandling of food during preparation in food service operations or in the home. Statistics collected by the Center for Disease Control and Prevention (CDC) provide a strong link between foodborne illness outbreaks and the end users in restaurants and retail. Consumers could be a source of fresh produce contamination in retail outlets.

Consumers touch fruit and vegetables as they make a decision on whether to purchase product. If a person's hands are contaminated because of improper hygiene, this product could be affected. The consumer may also place bare fruits and vegetables into shopping carts, which are generally unwashed. Cross-contamination by microorganisms present in the shopping cart onto fresh produce items could present a food safety hazard, especially if the cart was used to transport meat, fish, or poultry.

The final step of bagging produce items could present a food safety hazard if fruits and vegetables are placed in the same bag as raw meat or fish. When the consumer places grocery bags in the car, vehicle temperature and time to cooling determine the potential for present pathogens to multiply.

Retail operations and restaurants should take a proactive role in carrying out a risk assessment of their operations to understand the different types of hazards. When these potential problems are identified using the principles of HACCP, they may be controlled. Salads and juices are sometimes prepared in the fresh produce section of a retail operation. Workers should be trained in good food-handling practices to ensure food quality and safety are maintained:

- Ready-to-eat foods should be kept at safe temperatures.
- Food contact surfaces in food-preparation areas should be cleaned and sanitised regularly.
- Display counters and shelves should be cleaned regularly to prevent crosscontamination.
- Water used to wash produce should be potable, because poor-quality water is a major source of contamination.
- Colour-coded chopping boards should be used to distinguish boards for cutting produce from those used for cutting meat.
- Employees bagging product at the cashier's desk should be trained to avoid cross-contamination by placing raw meat in separate bags from fresh produce.
- Regular hand washing should be enforced to prevent product contamination.

All employees in a retail operation should thus have a basic knowledge of food safety and their responsibility of protecting the public. Outreach and education of the consumer would be the most effective means of explaining the dangers of mishandling food to the ultimate customers. Retailers may also assume the role of educating the public so the food-safety message reaches into people's homes.

Brochures, videos, and posters at display counters have been used effectively for consumer education in retail operations. Modern technology and a bit of creativity on the part of retailers could go even further to enlighten the public and promote safe handling of food. Restaurant chains represent one important point in the handling of fresh fruits and vegetables before consumption. Food-service operations have been linked to several foodborne outbreaks involving fresh produce.

Adequate training of food-service employees and implementation of a HACCP program to control food-safety in kitchens and food-display counters are the best ways to control food handling. The "Serve Safe" program developed by the International Food Safety Council of the National Restaurant Association Educational Foundation has become the industry standard in food safety training. Serve Safe is accepted in almost all U.S. jurisdictions that require employees' certification. The Serve Safe program provides accurate, up-to-date information for all levels of employees on all aspects of handling food, from receiving and storing to preparing and serving.

The International Food Safety Council promotes food-safety education to the restaurant and food service industry, and also conveys the industry's food-safety commitment to the public. The Serve Safe program has introduced a number of control measures for food-service operations similar to those mentioned above for retailers handling minimally processed product. In the United States, food service operations may also be subject to food-safety inspections, depending on the laws in each State. Only a very strict food-safety program in food-service operations would reduce the risk of pathogen contamination and illness to the ultimate consumer.

Consumer Handling

Fruit and vegetables, besides being perishable items, could be the source of mishandling by consumers, eventually leading to foodborne illness. Consumers sometimes mishandle produce by cross-contamination with meat items being placed in the same bag or cart. Temperature abuse in a hot car would promote the growth of microorganisms, if present. In the home, food-safety practices such as hand washing before handling fresh produce, or after handling meat, may not be observed.

Most consumers store produce in the refrigerator; however, some items are stored at room temperature. Room-temperature storage is good for tomatoes, bananas, and unripe climacteric fruits. However, refrigeration extends the freshness and slows bacterial growth if a produce item contains harmful microorganisms. The home refrigerator may also harbor harmful organisms if it is not cleaned routinely.

Meat drippings on shelves could cross-contaminate fresh produce placed on the same shelves. Consumers do not generally have a cleaning schedule for refrigerators, but this is a critical point in home hygiene in the kitchen. Refrigerator surfaces should be cleaned and sanitised to remove potential pathogens every two weeks, or more often, depending on contamination. Another reservoir for pathogenic microorganisms is the kitchen sink.

Consumers may place fresh produce items in the sink without washing or sanitising the area. This causes cross-contamination from items previously placed in the sink. Sinks should be cleaned and sanitised before being used to wash fresh fruits and vegetables. In addition, mixing of utensils would invite cross-contamination, leading to foodborne illness. Consumers may not always wash fruits and vegetables, but even the simplest washing with running water is sufficient to cause one $\log_{10}$ CFU/g reduction in microbes.

Fresh-cut fruit and vegetables should always be stored in refrigeration, first to extend shelf life, and then to reduce the potential growth of pathogens, which are usually mesophilic organisms. Chopping boards carry microorganisms in the form of biofilms if they are not cleaned and sanitised thoroughly. Consumers should implement good cleaning programs when using chopping boards for fresh produce, and preferably use different utensils than those used for meat.

The presence and survival of human pathogens in unpasteurised juices, sprouts, melons, lettuce, and berries is known. These commodities have been implicated in a number of foodborne illness outbreaks. Typical methods of washing and sanitising produce are not effective in removing pathogens, and increasing the knowledge of sources of microbial contamination requires more research.

Thus, scientists have tried to identify sources of human pathogen contamination and develop interventions to prevent contamination and remove or inactivate pathogens on fresh and minimally processed produce. These studies provide a foundation for the development of treatments to assure the microbiological safety of fresh produce.

However, experimental washing with hydrogen peroxide resulted in a 3-log (99.9 percent) reduction. Annous et al. demonstrated the inability of brush washing to decontaminate apples in field tests. It was thought that bacteria in inaccessible sites, growth of bacteria in skin punctures, and possible infiltration of bacteria into calyx and core tissues prevented their destruction. In work with apples, mature and immature fruit from various locations were examined for evidence of internal or external bacterial contamination.

A decaycausing fungus, Glomerella cingulata, permitted growth of E. coli O157:H7 in inoculated apples, probably because of a decrease in acidity that resulted from fungal growth. This clearly demonstrates the potential risk of using decayed apples for production of unpasteurised cider. Sapers et al. researched the microbiology of fresh and fresh-cut cantaloupe, providing knowledge about attachment and survival of bacteria, including surrogates of human pathogens on external melon surfaces. They studied the efficacy of washing treatments in decontaminating cantaloupe melons, and effective treatments for extending the shelf-life of fresh-cut cantaloupe.

Cantaloupe melons were artificially contaminated with nonpathogenic E. coli and Salmonella stanley, a human pathogen, and survival of these bacteria on melon rind during washing and their transfer to the melon flesh during fresh-cut processing was measured. Researchers found that chlorine and hydrogen peroxide solutions became progressively less effective in reducing contaminant populations as the time between contamination and washing increased. Surviving bacteria were transferred to the melon flesh and growth occurred.

Chlorine dioxide is one sanitiser approved by the U.S. Food and Drug Administration (FDA) and the

Environmental Protection Agency (EPA). Chlorine dioxide is said to be as powerful as peroxyacetic acid and more economical. It has less of a negative impact on the environment than quaternary ammonium salts, chlorine, or bromine. Ozone and peroxyacetic acid have been introduced to decrease the microbial content of fresh-cut produce. Chlorine bleach and bromine create carcinogenic trihalomethanes that enter the environment through drains.

Chlorine dioxide is also thought to destroy biofilms and prevent their formation. Chlorine continues to be the most widely used sanitiser in the produce industry in the United States; however, certain countries in the European Union have banned its use. Heat pasteurisation is one innovation receiving a great deal of attention in disinfecting fresh-cut apples and melons. The outside of the fruit is pasteurised before cutting, thus reducing the entry of surface microorganisms to the flesh. In fresh-cut processing operations, ATP bioluminescence is used to monitor food contact surface contamination.

It is a rapid means of informing sanitation crews if bacteria and organic material remain on food contact surfaces. This method of testing is, however, very expensive and finds application mainly in large processing operations. US Food Safety Programs: GAP, GMP, and HACCP. Use of GAP, GMP, and HACCP is voluntary in the fresh produce sector. However, retailers have demanded that their suppliers implement structured food-safety programs under various schemes and these practices are now subject to auditing by a third party.

Good Agricultural Practices (GAPs) describe preventive measures implemented in farming operations to reduce product contamination. The U.S. government in 1998 introduced a "Guide to Minimise the Microbial

Contamination of Fresh Fruit and Vegetables". The major risk factors identified as needing control included water, manure, worker health and hygiene, sanitary facilities, pest control, and traceback. GAPs provide guidance for food-safety practices in the field.

Implementing GAPs provides some assurance to the retailer that product is safe from the farmer's gate. A small number of foodborne illnesses have been associated with contamination at the farm through an ill harvester, or use of unsanitary water. Current GMPs were made law in U.S. food-processing and handling facilities and this code has been reviewed. The code provides guidance on the safe handling of product in buildings used for food production. In the fresh produce sector, packhouses should follow guidance set out in cGMPs in order to put out wholesome food product.

The principles of HACCP, a system developed by the NASA space program in the 1950s, are now used as the basis for risk assessment and initiating food-safety measures throughout the industry. Although most fresh produce farming and packing operations do not have critical control points, the HACCP program forces operations to go through each area in a packhouse, conducting a risk assessment. The assessment may also indicate where GMPs are failing and help in improving GMPs. The use of GAPs, GMPs, and HACCP in the fresh fruit and vegetable industry provide the basic framework for safe products for the consumer.

Distribution and Retail

To aid the integration of the supply chain from the source to the end consumer, automated systems are being developed. These data-based systems would provide complete and constant visibility of product, package, purchasing, and distribution of food. Automated warehouse management systems using robotics could increase accuracy of the picking and assembling of pallets for dispatch. With laws restricting the weight of product lifted at food plants, automated systems could also aid manufacturers in compliance with labor laws. As customer demands for fresh product increases, smaller, more frequent deliveries will be required.

Automated warehouse management systems would again improve accuracy of product rotation and picking for small deliveries. Automated systems would improve inventory control and real-time tracking of products that pass through a warehouse. The advent of barcoding has revolutionised the way product movement is managed. A unique barcode is prepared for product harvested on the farm. This barcode stores information on the grower, farm lot number, day harvested, and crop harvested. A barcode placed on a bin of fruit after harvest could then be scanned at the packhouse at intake. This code stays with the bin until it is repacked into consumer size items.

If product were stored before repackaging, the warehouse management system would assist in stock rotation based on information received from the barcode. When product is repacked into consumer size packages, a new barcode is placed on the product to assist in tracing finished product back to the field in case of an emergency. The Uniform Code Council (UCC) takes a global leadership role in establishing and promoting multi-industry standards for product identification and related electronic communication.

The purpose of CC/EAN-128 is to establish a standard way of labeling a package with more information than just a product code. It provides supplemental information such as batch number and "use before" dates. The symbology for this is the UCC/EAN-128, which can be scanned to capture automatically the details related to the location. International standards in barcoding include the use of a Global Location Number or GLN. This is a number that identifies any legal, functional, or physical location within a business or organisational entity such as

- Legal entities—whole companies, subsidiaries, or divisions such as supplier, customer, bank, forwarder;
- Functional entities—a specific department within a legal entity, such as the accounting department;
- Physical entities—a particular room in a building, such as a warehouse or warehouse gate, delivery point, or transmission point.

Such GLNs are governed by strict rules to guarantee that each one is unique worldwide. They always have the following features:

- They are numeric.
- They have a fixed length of 13 digits, which must be processed in their entirety.
- They end with a check digit.

The GLNs can be presented in barcode format and physically marked onto trade units to identify the parties involved in the transaction (buyer, supplier), for example (1) transport units (consignor, consignee), (2) physical locations (place of delivery, place of departure). X-ray technology is another method with great potential in identifying physical contamination in fresh fruits and vegetables. The widely used metal detectors are not able to detect wood chips, plastic, and other nonmetal contaminants. Thus, machine vision systems could help in eliminating foreign contaminants in food.

The cost of purchasing equipment and adapting it to fresh-produce operations far exceeds the benefits from using this technology. Although some in the produce industry are still grappling with the introduction of barcoding systems for traceability programs, other sectors in the food industry have moved on to the new technology of radio-frequency identification (RFID) systems, a technology that uses grain-sized computer chips to track items at a distance. Each tiny chip is connected to an antenna that picks up electromagnetic energy beamed at it from a reader device.

When it picks up the energy, the chip sends back its unique identification number to the reader device, allowing the item to be remotely identified. "Spy chips," as they are sometimes called, can beam back information anywhere from a few inches to 30 ft away. This RFID will be used in the future to trace product. Retailers such as Walmart have started demanding this technology for packaged goods. Within time, this technology could be used to assist in stock rotation and distribution of fresh produce. Radio frequency identification is not necessarily better than barcodes.

The two are different technologies and have different applications, which sometimes overlap. The big difference between the two is that barcodes are a line-of-sight technology. In other words, a scanner has to see the barcode to read it. This means that people usually have to orient the barcode towards a scanner for it to be read. Radio frequency identification, by contrast, does not require line of sight. RFID tags can be read as long as they are within range of a reader. All existing RFID systems use proprietary technology, but many of the benefits of tracking items come from tracking them as they move from one company to another and even one country to another.

Traceability for the Industry

Traceability is defined as the "ability to trace the history, application or location of that which is under consideration." The Codex Alimentarius Commission of the Food and Agriculture Organisation defines traceback is the ability to track product from the consumer point of purchase back to the grower. The objectives of a good traceability system should be to improve supply management and aid traceback of food-safety and quality parameters. Some companies use their traceability programs as a marketing tool to differentiate them from competitors because of the benefits of the program.

In most cases, a well-run traceability program can generate revenue for a company because it improves efficiency at all levels. The source of a safety or quality problem can be isolated and managed using a detailed traceability program. Benefits of traceability also include the fact that unsafe products are not shipped, and the risk of bad publicity, lawsuits, and recalls is reduced. A detailed tracing system would speed corrective actions when food-safety and quality problems are identified.

Many large restaurant chains and retailers now expect their suppliers to create traceability systems. These systems are verified through third-party audits of overall food-safety systems. If a product recall or foodborne illness occurs, an investigation is well controlled if a company has a means of retrieving information. This would help in identifying a problem in the production system and point to immediate corrective actions. Under the FDA Recording and Reporting Rule of the Bioterrorism Act 2002, food companies must provide the FDA with source data, including immediate previous source (IPS) and immediate subsequent recipient (ISR) for every component of a food item.

U.S. and international food processors, transportation, and distribution companies must keep records if food or biological products are imported or exported from the United States. Companies have developed innovative software databases in information systems with the ability to store detailed crop history and distribution data. Guidelines have to be set so that meaningful data can be stored and retrieved when needed. Systems usually include information on the shipper, date harvested, field, and picker. The basic requirements of information databases are

- To quickly isolate affected product in the event of a recall;
- To track spray intervals to harvest and pesticide application records, if pesticide use is a concern;
- To keep historical data on fields where crops were grown.

The U.S. Government has developed systems to monitor the incidence of foodborne illness. FoodNet (www.cdc.gov/foodnet/) and PulseNet (http://www.cdc.gov/pulsenet/) are surveillance systems that increase the capability of the entire food supply chain to respond to food-safety problems before more consumers are affected. FoodNet—Foodborne Diseases Active Surveillance Network (FoodNet)—is a collaborative project of the CDC, ten Emerging Infections Program (EIP) sites, and the U.S. Department of Agriculture.

PulseNet is a national network of public health and food regulatory agency laboratories coordinated by the Centers for Disease Control and Prevention (CDC). The network consists of state health departments, local health departments, and federal agencies (CDC, USDA/ FSIS, and FDA). PulseNet participants perform standardised molecular subtyping (or "fingerprinting") of foodborne disease-causing bacteria by pulsed-field gel electrophoresis (PFGE), which can be used to distinguish strains of organisms of public health significance.

Scientists can trace fresh produce microbial contaminants using genetic data. The global food-safety testing market is expected to reach as much as $415.6 million by the year 2009. The increase in pathogen test sales continues to rise because of many factors, including

- Consumer demand for safe food;
- Evolution of bacteria and emergence of new strains;
- Increased government regulation.

Food testing globally focuses on pathogens, pesticide residues, genetically modified organisms (GMOs), and other contaminants. In the United States, pathogen testing far exceeds pesticide testing in foods. This is different in the European Union, where testing focuses on pesticide residues. The European Union has also required increased documentation of GMO-containing foods and testing for identity preservation may increase in the future.

Customers are reassured when information supplied by companies is verified by third-party certification. Different forms of auditing include self-audits, secondparty audits, and third-party audits. Self-audits are used by companies to monitor the efficiencies of in-house programs: food safety, GMP, HACCP, and so on. Self-audits may be used as a tool to monitor execution of QA and QC programs. One setback would be the familiarity of the auditor with the operation and his inability to view errors in the system objectively.

A company verifying practices of its suppliers conducts second-party audits. Some bias is removed, because the company would want the best-quality product as raw material and packaging from vendors. Third-party audits are conducted by neutral auditing firms that carry out inspections on behalf of a retailer or other party of interest. The auditing firm does not usually have a stake in the business and so the assessment is felt to be unbiased. In order to be effective, third-party auditors should be knowledgeable about the business they are auditing and well versed in the standard against which they are auditing.

Different auditing and certification protocols exist in different parts of the world. Each protocol may be applied in a specific area of the food chain, but there is no one correct auditing scheme. Nations have not yet united all schemes under one accreditation body. Good Agricultural Practices (GAPs) refer to food-safety guidelines for field operations and each auditing firm has its own verification system for monitoring GAPs. Another scheme for farm safety, EurepGAP, has been developed in Europe. This protocol provides many allowances for environmental and social issues in the growing operation.

European retailers who helped develop the EurepGAP auditing scheme ask growers and shippers to subscribe to this auditing program. Good Manufacturing Practices (GMPs) have been clearly defined by the U.S. FDA for food processing operations. Different auditing firms in the United States would interpret GMP guidelines and audit packhouses and processing plants based on their own checklist. The U.K. Food Safety Act 1990 states that all persons involved in supplying food have an obligation to take all reasonable precautions and exercise all due diligence in avoiding failure when presenting food products to the consumer. Retailers are thus obligated to verify technical performance at food-production sites.

In 2003, the British Retail Consortium (BRC) proposed a Food Technical Standard applying to enclosed areas for food production. The standard has been updated with version 4, the BRC Global Standard—Food, in June 2005. This standard for food establishments is seen to be a major international standard specifying safety, quality, and operational criteria within a food manufacturing operation. Accredited auditing bodies inspect food establishments against this standard. Schemes such as the BRC's Global Food Standard provide training for auditors based on the set standard. Training ensures uniformity in auditing.

Another example of a certification protocol is the International Organisation for Standardisation's ISO 9000 series, or "EN 29000," in Europe. In 1987, ISO published a series of five international standards (ISO 9000, 9001, 9002, 9003, and 9004), developed by ISO Technical Committee (TC) 176 on quality systems. This series, together with the terminology and definitions contained in ISO Standard 8402, provides guidance on the selection of an appropriate quality management program (system) for supplier operations.

Conformance to ISO 9000 standards is required in purchasing specifications by some companies. The ISO 9000 Standard Series was adopted in the United States as the ANSI/American Society for Quality Control (ANSI/ASQC Q 9000 series). In Europe, the series was adopted by the European Committee for Standardisation (CEN) and the European Committee for Electrotechnical Standardisation (CENELEC) as the European Norm (EN) 29000 Series. Many countries have national standards that are identical or equivalent to the ISO 9000 Standard Series.

Growers use results of third-party audits of GAPs as a tool to demonstrate that effective due diligence programs are in place. When the FDA introduced GAPs for the produce industry, auditing firms used these guidelines to develop extensive auditing programs for the fruit and vegetable industry. Retailers rely on auditing firms to provide them with the assurance they need from their suppliers. However, fruit and vegetable growers in the United States have the following concerns:

- There is no government regulation of third-party auditing companies.
- Food auditors from the food-processing industry attempt to apply the same standards of auditing in open fields.
- Cost of auditing fees is seldom supported by an increase in revenue for growers.
- Different retailers may demand different auditing programs, because there is no industry standard.

A major retailer, Tesco, has developed the Tesco Nature's Choice standard for growing and packing operations. Walmart has developed the Walmart Global Procurement Group Audits. Safe Quality Food (SQF 1000), a scheme owned by the U.S. Food Marketing Institute, verifies food safety as well as quality of product. Retailer auditing programs are not necessarily similar and may increase costs for those growers that supply different stores. Microbial food safety is the focus of United States based produce industry audits.

However, there are insufficient data to establish actual risks of microbial contamination in the industry. Thus, auditing is based on many assumptions of risk. Even companies that passed audits have been associated with foodborne outbreaks. This shows that foodsafety programs allow for a minimisation of risk, but not complete elimination of microbial contamination. The consolidation and globalisation of the retail industry has brought a new dimension to the auditing process. It is hoped that the market will consolidate its basic food safety requirements. The U.S. culture of regulatory control of the food industry is different to the European Union, where a culture of private certification exists.

In addition, niche markets may make higher demands of suppliers. Auditing schemes could be made more effective by the following steps:

- Developing data to assist in objectivity of decision-making in auditing;
- Testing and evaluating management tools to add objectivity to audits;
- Introducing controls on accreditation schemes for auditors in all countries;
- Harmonising international auditing schemes so they cover major quality and food-safety concerns;
- Developing product-specific and region-specific auditing schemes;
- Introducing an international accreditation body using set standards to verify programs of all auditing bodies, thus making the auditing process uniform.

The Global Food Safety Initiative (GFSI) was created in May 2000 at the initiative of a group of major retailers, including METRO AG, Rewe, Edeka, and Globus. The GFSI is a working group within the CIES (The Food Business Forum), an international network of companies in food retailing and food production. In total, 250 major food retail companies are represented in CIES. GFSI is a separate legal entity governed by an Advisory Group. The mission of GFSI is to strengthen consumer confidence in the food bought in retail outlets.

GFSI is to develop product-related standards for food safety that cover all links of the food chain. It wants all suppliers to be audited and certified according to a uniform international standard. This will be achieved by implementing and maintaining a scheme to benchmark food-safety standards (for private label products) as well as farm assurance standards. GFSI will facilitate mutual recognition between standard owners and ensure worldwide integrity in the quality and the accreditation of food safety auditors.

The retail industry drives the introduction of GHPs in the food industry. As the last point of contact in the fresh-produce chain, retailers could play a proactive role in educating the public on safe food-handling practices. Programs retailers may use to educate the public include the following:

- Use of membership cards and frequent shopper cards to store data on buying patterns of regular customers;
- Tailor-made food safety information to send to customers;
- Posting signs on the sides or handles of shopping carts to deliver food-safety information to the customer before they enter the store;
- Posting point-of-sale signs, wall hangings, and floor markers;
- Distributing fliers with information in shopping bags to keep the customer informed and aware of food-safety issues concerning the item purchased;

- Posting food-safety information on retailer websites;
- Sponsoring radio/TV and other media programs promoting food safety;
- Using videos in store to impart knowledge of safe food-handling practices;
- Using demonstration stations in the store to sample new food items while incorporating food-safety education.

In the United States, Federal Governmental agencies (the FDA, USDA, and CDC) are responsible for setting and monitoring food-safety standards, as well as communicating key food-safety knowledge to consumers. Local government agencies have been busy educating and informing consumers in the areas of food safety. "Fight Bac!" is one government program that was a success and can serve as a model for conveying food-safety knowledge to the consumer. Educational material for consumers must be presented in a convenient way.

Consumers who were interested in receiving information found brochures in supermarkets informative. Consumers also preferred it if information was presented in easy-to-follow picture form, instead of wordy material. Information should also be targeted toward children as part of school studies and with special activities. For example, safe food handling could be a required component of any social studies curriculum. It is important to channel food-safety education to school-aged children, the consumers of the future. Educating children in safe food-handling practices would perhaps reduce foodborne illnesses for future consumers.

Role of Surrogates and Indicators

FDA/CFSAN (2001b) defines an indicator organism as "a

microorganism or group of microorganisms that indicate a food has been exposed to conditions posing an increased risk that food may be contaminated with a pathogen or held under conditions conducive for pathogen growth." Indicator analyses are sometimes used to determine the effectiveness of sanitation programs. An indicator could be a specific microorganism, a metabolite, fragment of DNA, or other substance.

The indicator should be present in the food when the target pathogen is present, should have similar growth patterns as the target, and be easy to detect and quantify. Surrogate organisms are a unique type of indicator. They mimic growth and survival patterns of a pathogen and can help in studying what occurs with a pathogen during handling and storage. Surrogates are used to study the effects and responses to specific processing treatments. In the fresh fruit and vegetable industry, surrogates could be used to determine the effectiveness of a produce wash. They are usually prepared in a laboratory and introduced onto the produce item. They may also be microorganisms occurring naturally on the produce item.

Indicators are naturally occurring, whereas surrogates are introduced artificially on the food item for the study. Surrogates are frequently used for research as opposed to pathogens, because scientists need to prevent the release of harmful organisms in a production environment or, in the case of fresh produce, into a field. However, isolated pilot facilities have used actual pathogens for controlled research. Mitigation measures in the fresh produce chain are mostly based on risk assessment and making an educated guess as to what measures would reduce the risk.

Indicators and surrogates could be used to assure safety for individual produce items in all stages of the fresh produce chain: growing, harvesting, processing, handling, storage, distribution, and retail. These microorganisms may be used to determine the effectiveness of a water treatment or composting on pathogen reduction. If an indicator is in low concentration or not present, this could mean a treatment would be effective in reducing growth of the target organism. It is difficult to select the right target organisms, as well as an indicator organism that would relate to the target.

A surrogate should have the following main characteristics:

- It should be a nonpathogenic strain of the target organism.
- It should demonstrate similar behaviour as the target organisms in the environment studied.
- It should be easy to prepare and stable.
- It should exhibit enumeration, sensitivity, and inexpensive detection means.
- It should attach to the produce surface at a similar rate as the target organism.
- It should exhibit genetic stability to allow the test to be reproduced independently.

The use of surrogate organisms started in the low-acid canning industry. The thermophilic organism Bacillus stearothermophilus was used as a surrogate to establish processing conditions to destroy Clostridium botulinum spores. Gram-positive and gram-negative bacteria, viruses, and protozoan cysts are pathogens associated with produce and foodborne outbreaks. Strains of Listeria innocua have served as nonpathogenic surrogates of the pathogen Listeria monoytogenes.

Use of indicators and surrogates can be the most scientific means of generating data to validate the effectiveness of microbial control measures. Additional research is needed to identify specific surrogates and indicators for ready-to-eat and fresh-cut produce, thus helping to validate decontamination steps in the process. It is unknown if pathogens would develop increased resistance as a stress response when subjected to intervention steps. In addition, the country of origin and packing environment may induce a stress response.

Protocols have to be developed to accurately measure and retrieve indicator microorganisms after a stress is introduced. For an extensive list of research needs in the use of indicators and surrogates, the reader is referred to FDA/CFSAN. Scientists raise concerns that researchers are not using established or comparative studies to validate their surrogates or indicator strains with the appropriate pathogen. The number of published works on attenuated pathogens has increased in several laboratories. However, researchers presume that elimination of virulence traits or entire plasmids has no impact on stress tolerance, phenotypes, competitiveness, or environmental persistence.

Challenges for the Fresh Produce Industry

Consumers now benefit from worldwide trade with year-round supplies, good quality, lower prices, and a wide variety of food. Food safety standards may sometimes present technical barriers to trade, but improvements in standards are likely to extend worldwide with the spread of private and governmental control. In the past, foodborne outbreaks were localised, but now, with increase in international trade, low-level contamination could possibly spread to many regions and even cross borders.

Trade volumes are increasing for fresh, minimally processed, or high-value food, with an increase in imports from developing countries needed to fill the gaps. No scientific evidence has shown imported fresh produce to pose any greater food safety threat than U.S. grown products. In addition, less developed countries are acquiring higher food-safety standards to assist in entering new markets. Both Government agencies and private retailers and importers are driving this effort to increase food-safety standards.

The higher standards do not, however, give complete protection from disease or injury. Branded products, if perceived to be safe, could improve a firm's image and international competitiveness. Private retailers and importers are strongly motivated to prevent a food-safety crisis, because their companies may suffer from loss of reputation, plants closing for cleanup, reduced stock prices, lawsuits, or higher insurance premiums.

Large organisations are now promoting vertical integration, controlling commodities at two or more stages of production. Private companies have promoted the following:

- Self-regulation;
- Vertical interaction;
- Third-party certification;
- Use of HACCP principles in risk assessment;
- Application of voluntary GAPs in growing operations.

Government regulation also plays an important role in addition to the private sector. Industrialised nations saw the development of new food-safety regulation in the 1990s with improvements in science and increased awareness of foodsafety risks. Regulatory agencies in industrialised nations were seen to follow main trends, forming agencies to focus only on food safety. Risk analysis was used to design guidelines for the industry and a "farm-to-table" approach has been used to address food-safety hazards.

Some government agencies have adopted a HACCP-based system for new regulation of microbial pathogens in food. Newly identified hazards have encouraged adoption of more stringent standards and market performance has improved through the government providing the industry with more food safety information. Trade disputes over food safety may be persistent, but are manageable. Food safety challenges vary with each commodity. Concerns differ, depending on the perishability of the product, the nature of human health risks, the possible link of the food-safety issue with productivity, and the extent of vertical coordination and cooperation among stakeholders.

Consumers are more frequently eating at food-service establishments. Even foodproduction practices have changed, leading to a host of new food-safety risks for the industry to manage. A marked increased in cut, ready-to-eat fruit and vegetables increases the challenge of preparing safe food. Consumers are demanding convenience food items to suit the hectic pace of the life we now lead. Washing, cutting, and preparing noncooked vegetables increase the potential for product contamination. This could increase the potential for more foodborne outbreaks because food handling is increased.

Introduction of mandatory HACCP in juice processing in the United States was one move of the government to control food safety in the sector. Catering operations are monitored by food inspectors, whereas fresh-cut processing operations are audited at the request of retailers and restaurant chains. Food-safety programs in fresh-cut operations, however, are voluntary and not under mandatory regulation by government bodies. Food-safety disputes could arise in trade with the appearance of new hazards or increase in volumes from new sources.

Rising standards or changing food-safety regulation in industrialised countries would create new challenges for trade from developing nations. Different countries would have unique ideas with respect to consumer risk preferences. Nonscience issues also play a role in regulatory decisionmaking and this would vary from one country to the next. Growth in demand and increased access to markets will continue to expand food trade. Improvements in food safety are essential to consumer welfare and both producers and consumers have a stake in seeing the industry make great strides in this area.

3

Transportation of Fruits and Vegetables

Fruits and vegetables not only provide human beings with nutritional and healthy foods, but also generate a considerable cash income for growers in many countries. However, they typically have a high moisture content, tender texture, and high perishability. If not handled properly, a high-value nutritious product can deteriorate and rot in a matter of days or even hours. Therefore, a series of sophisticated technologies have been developed and applied in post-harvest handling of horticultural crops in the last few decades.

The quality of fresh fruits and vegetables has a decisive effect on their value. This is particularly true when consumers have a high income and the market provides a wide choice of produce. It is not uncommon in markets for high-quality fruit to have a price two to three times higher than mediocre produce of the same kind. While it is cultural practices which determine the quality of produce at harvest, proper handling ensures that the quality is preserved until the produce reaches the consumer. Quality assurance is a prerequisite for high-value fresh fruits and vegetables.

Post-harvest Losses

Poor planning or inaccurate production and market information may lead to over-production of certain fruits or vegetables which cannot be sold in time. This situation occurs most frequently in areas where transportation and storage facilities are inadequate. Produce may lie rotting in production areas, if farmers are unable to transport it to people who need it in distant locations.

Although many people believe that postharvest losses are higher in less industrialised countries, this generalisation may not be always true. Higher losses may occur in developing countries for lack of good facilities and technologies. However, post-harvest losses may be lower in less urbanised developing countries, where the products need to be transported a shorter distance to market, and there is a shorter time lag period between harvesting and consumption. Consumers in such countries are also more likely to buy lower-grade products. In industrialised countries, on the other hand, produce is shipped for longer distances, the marketing system is more complicated, and consumers demand good quality when they buy food.

All these factors contribute to greater postharvest losses which may not be fully compensated by better facilities and technologies. For instance, Bourne cited a survey report that losses of produce after arrival in New York City were 1.7 14.2%. This includes only losses in wholesale markets, retail outlets and by consumers. It is safe to say that post-harvest losses occur in every country, but the magnitude and major causes of losses and the effective remedial methods differ greatly from one country to another. In order to solve specific problems in specific areas effectively and economically, a comprehensive knowledge of the nature of post-harvest losses, a grasp of various kinds of technologies, and a tactical selection of strategies is necessary.

Post-harvest losses have many different forms. The most obvious loss is a quantitative loss, such as reduced weight and partial or total waste of a product due to decay or senescence. A less discernible loss is a qualitative one, such as deterioration in texture, flavor or nutritional value. Other forms include loss of viability of propagules and reduction in monetary value due to reduced prices.

Post-harvest losses can occur in the field, in packing areas, in storage, during transportation, and in wholesale or retail markets. Severe losses occur because of poor facilities, lack of know-how, poor management, market dysfunction, or simply the carelessness of farmers or workers. Postharvest losses may also occur at consumers' homes, in the kitchen or on dining tables. However, losses after produce has left the retail market are generally difficult to control by agricultural means.

The causes of post-harvest losses are:

— *Mechanical Injury.* Owing to their tender texture and high moisture content, fresh fruits and vegetables are very susceptible to mechanical injury. Poor handling, unsuitable containers, improper packaging and transportation can easily cause bruising, cutting, breaking, impact wounding, and other forms of injury.

— *Physiological deterioration.* Fruit and vegetable cells are still alive after harvest, and continue their physiological activity. Physiological disorders may occur due to mineral deficiency, low or high temperature injury, or undesirable atmospheric conditions, such as high humidity. Physiological deterioration can also occur spontaneously by enzymatic action leading to overripeness and senescence, a simple aging phenomenon.

— *Parasitic diseases.* High post-harvest losses are caused by the invasion of fungi, bacteria, insects and other organisms. Microorganisms attack fresh produce easily and spread quickly, because the produce does not have much of a natural defense mechanism, and has plenty of nutrients and moisture to support microbial growth. Post-harvest decay control is becoming a more difficult task, because the number of pesticides available is falling rapidly as consumer concern for food safety increases.

Principles of Transportating Horticultural Crops

Transportation is often the most costly factor in the marketing channel, and for airfreighted export crops the cost of transportation may exceed the cost of production. The method of transportation for fresh fruit and vegetables is determined by distance, perishability and the value of the product. Whatever the method used, the principles of transport are the same:

— Loading and unloading should be as careful as possible.

— Transmit times should be as short as possible.

— The product should be well protected in relation to its susceptibility to physical injury.

— Jolting and movement should be reduced as much as possible.

— Overheating should be avoided.

— Water loss by the produce should be restricted.

— The required conditions of preservation should be obtained and maintained constantly, regarding in particular temperature, relative humidity and air circulation.

Handling and Moving Short Distances

Marketing and physical distribution of fresh produce inherently means moving the produce. The commodities are handled, either manually or mechanically, many times from harvest and through the distribution process before the consumer buys and prepares them to eat.

Handling operations are seldom given much thought by the individuals directly involved in moving the process, particularly when the produce is only moved short distances. Figure 1 gives an indication of the number of handling steps that will probably be endured by the produce.

Notice that most of them are movements over very short distances and most probably by direct manual handling. These short moves are usually highly repetitive operations and unless the personnel concerned are properly trained and motivated then the physical handling is likely to be very bad leading to produce injury, spoilage and high losses.

Figure 2 shows some useful equipment for moving produce short distances which would not only make the job much easier for the workers, it would also speed up operations and reduce physical injury to the produce at the same time.

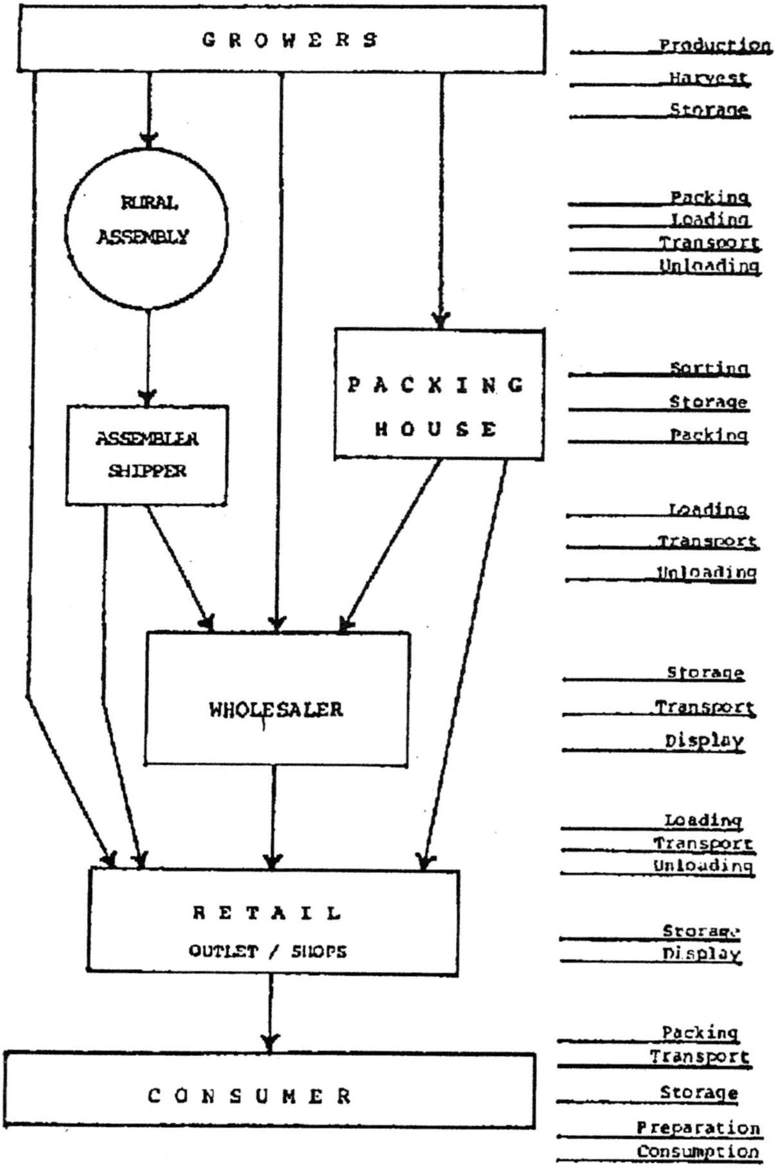

Figure 1. Handling steps during fresh produce marketing

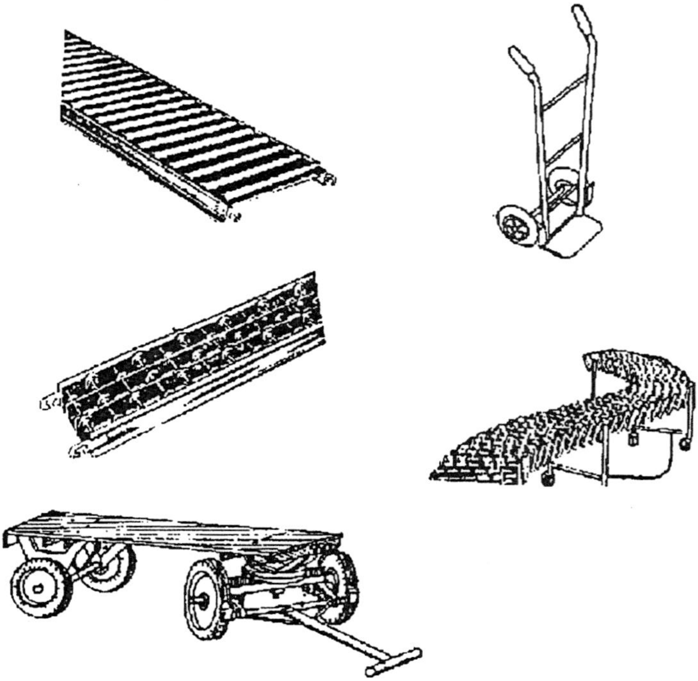

Figure 2. Equipment for moving fresh produce short distances

Roller-conveyors, be they motorised or gravity fed are of great assistance in the packing house for moving boxed produce but are equally useful for loading and unloading of pickups, trucks and stores. Hand-trucks can carry up to six or more crates of produce, are very manoeuvrable light and durable, and do not cost a lot of money. Hand-carts with the front wheels steered by the tow-bar can carry a lot more produce but need to be used on more level ground. Pallet-trucks are in everyday use in the Eastern Caribbean but are not yet used with any regularity for fresh produce handling and movement, probably because packaging has yet to be sufficiently standardised to benefit from pallet operations.

Handling and Stowage during Transportation

Dropping of packages during loading and unloading is a frequent cause of damage to the produce and to the package, but can be minimised by:

- — Using pack weights and designs which are compatible with the handling method.
- — Correct supervision and management of loading and unloading to prevent careless handling and to ensure workers are strong enough and tall enough for the Job.
- — Using ramped loading bays gives a tremendous advantage when loading trucks with produce.
- — Providing shelter from sun and rain at loading and unloading areas.
- — Using trolleys, conveyors and fork-lift trucks to reduce the amount of manual handling.

The method of stowage of the produce in the transport vehicle will depend on the pack, the commodity and the type and size of the vehicle but should always be carefully planned and managed to minimise both physical and environmental damage. The following are some useful guidelines:

- — Loading tightly to reduce movement and make best use of space.
- — Distribute weight evenly.
- — In mixed commodity or dispatch loads, stow goods in reverse order to their unloading sequence.
- — Provides gaps for ventilation.
- — Only stack to a load height which the lowest containers can withstand without crushing or damage to the pack.

— Do not exceed the capacity of the vehicle.

— Ensure the vehicle is properly maintained, breakdowns are time delaying and may lead to excessive or total spoilage of the produce.

— Be selective as regards the vehicle operator, e.g. band and inexperienced drivers will mean more damage to the produce and to the vehicle.

Road Transportation

Inland transportation of horticultural crops is usually by rail or way truck. Overseas transportation is by sea or by air. A limited amount of high-valued produce is sometimes transported overland by air. The basic requirements for conditions during transportation are similar to those needed for storage, including proper control of temperature and humidity and adequate ventilation. In addition, the produce should be immobilised by proper packaging and stacking, to avoid excessive movement or vibration. Vibration and impact during transportation may cause severe bruising or other types of mechanical injury.

CA (Controlled Atmosphere) containers are not used much for transporting fruit and vegetables. Only a small number of CA containers are used for long-distance shipping of highly perishable, high-value products in the international trade. Refrigerated containers and trailers are more often used for longdistance shipping, whether by sea, rail or truck.

Shipping by refrigerated trucks is not only convenient, but also effective in preserving the quality of product. However, both the initial investment and the operating costs are very high. Another possibility is insulated or properly ventilated trailer trucks. Precooled products can be transported by well insulated non-refrigerated trucks for up to several hours without any significant rise in product temperature.

There are considerable cost savings without any sacrifice of quality if trucks are only insulated, rather than refrigerated, for short-distance shipping. If the product is not precooled or if the shipping distance is long, a ventilated truck is a better choice than an insulated truck without ventilation and without refrigeration. Ventilation alone does not usually provide a uniform cool temperature, but it may help dissipate excessive field heat and respiration heat, and thus avoid high temperature injury.

Temperature management is critical during long distance transport, so loads must be stacked to enable proper air circulation to carry away heat from the produce itself as well as incoming heat from the atmosphere and off the road. Transport vehicles should be well insulated to maintain cool environments for pre-cooled commodities and well ventilated to allow air movement through the produce.

During transport, produce must be stacked in ways that minimise damage, then be braced and secured. An open air vehicle can be loaded in such a way that air can pass through the load, and provide some cooling of the produce as the vehicle moves. Traveling during the night and early morning can reduce the heat load on a vehicle that is transporting produce.

Mixed loads can be a serious concern when temperature optima are not compatible (for example, when transporting chilling sensitive fruits with commodities that require very low temperatures) or when ethylene producing commodities and ethylene sensitive commodities are transported together. High ethylene producers (such as ripe bananas, apples, cantaloupe) can induce physiological disorders and/or undesirable changes in color, flavor and texture in ethylene sensitive commodities (such as lettuce, cucumbers, carrots, potatoes, sweet potatoes).

For domestic transportation the use of road vehicles offers substantial advantages of convenience, availability, flexibility permitting door-to-door delivery, and reasonable cost of transport. The use of road transportation for fresh produce is increasing and likely to increase in countries all over the world. Produce may be transported by pick-up, enclosed truck, open truck or refrigerated vehicle.

Enclosed Vehicle Transportation

Enclosed vehicle these re only suitable for short journeys, unless provided with a cooling system, since the produce inside heats up rapidly. However, they protect the produce from pilferage and physical injury, and are often used for urban retail delivery.

Open Vehicle Transportation

Pick-ups and open trucks are the commonest type of road transport. They are often fitted, before or after purchase, with frames to ease stacking and covering. Natural ventilation is usually sufficient to prevent overheating of produce over relatively short journeys and the most versatile types have a fixed roof and tarpaulin drapes which can be pulled along the sides and back to allow access for loading and unloading at any point. These loose awnings are not in contact with the produce and so allow for ventilation systems on these types of vehicles are unnecessary for short journeys, but when transit time are more than a few hours, adjustable louvres and air-scoops may be needed.

Bulk loads

Bulk loads of produce should be carefully loaded so as not to cause mechanical damage. Vehicles can be padded or lined with a thick layer of straw. Woven mats or sacks can be used in the beds of small vehicles. Other loads should not be placed on top of the bulk commodity.

Figure 3. Bulk loads

Cooling open loads is desirable whenever possible. A truck ventilating device can be constructed for an unrefrigerated open vehicle by covering the load loosely with canvas and fashioning a wind catcher from sheet metal. Scoop should be mounted at the front of the bed and should reach somewhat higher the height of the cab.

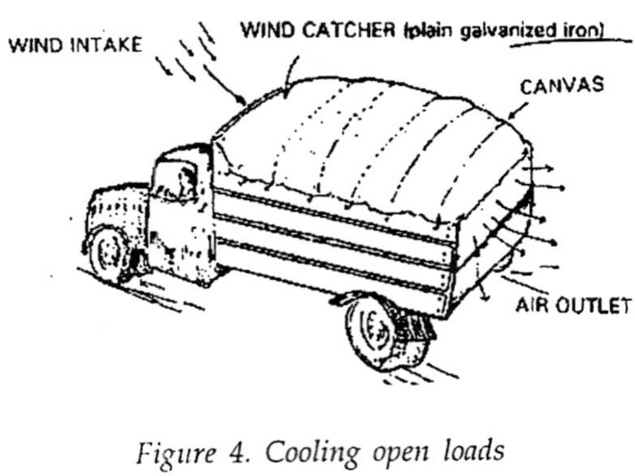

Figure 4. Cooling open loads

High transportation speeds and/or long distance transport run the risk of causing excess drying of the crop.

This ventilating system was designed for hauling bulk loads of fresh fava beans in Iran. The wind catcher and ducts were constructed using wooden crates. After removing their end panels the crates were wired together into the pattern shown below.

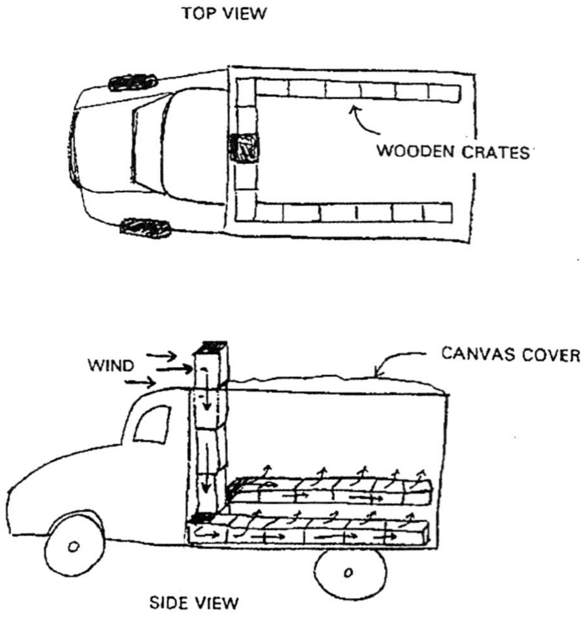

Figure 5. Open vehicle ventilating system

Air flows upward through the load during transport, helping to keep the produce from overheating. This system has also been used in pick-up trucks, and for hauling bulk greens and green beans. Best results are obtained when transporting during the early morning hours, before sunrise.

Refrigerated Vehicle Transportation

Some highly perishable products may justify the use of refrigerated vehicles. Ice is not generally used to refrigerated trucks because of weight and corrosion constraints and for most refrigerated vehicles a mechanical system is used. Truck-based mechanical refrigeration systems vary in their cooling capacity. Most are only suitable for maintaining the temperature of produce which has been pre-cooled by other means, and they possess fairly weak air-circulation fans sufficient only to allow refrigeration of air heated up by low respiration of the cool produce. Some form of ventilation may be necessary on longer journeys to prevent oxygen depletion and carbon dioxide accumulation.

A few refrigerated vehicle types, usually reefer containers mounted on the back of flatbed trucks, are capable of rapid forced-draught cooling of warm produce, but these are generally an exception because of their expense.

For optimum transport temperature management, refrigerated trailers need insulation, a high capacity refrigeration unit and fan, and an air delivery duct. The figure 6 shows these and other desired features in a top-air delivery trailer. The condition of the inside of a refrigerated trailer affects its ability to maintain desired temperatures during transport.

There is often a tendency to use the relatively low capacity refrigerated trucks as the precooling system for export produce. The trucks were not designed as pre-coolers and the results are not satisfactory. In other instances, these same trucks may be used as mobile refrigerators to be installed for days or weeks at a time, as a form of refrigerated store. Again the results are far from satisfactory and spoilage levels are very high. In addition,

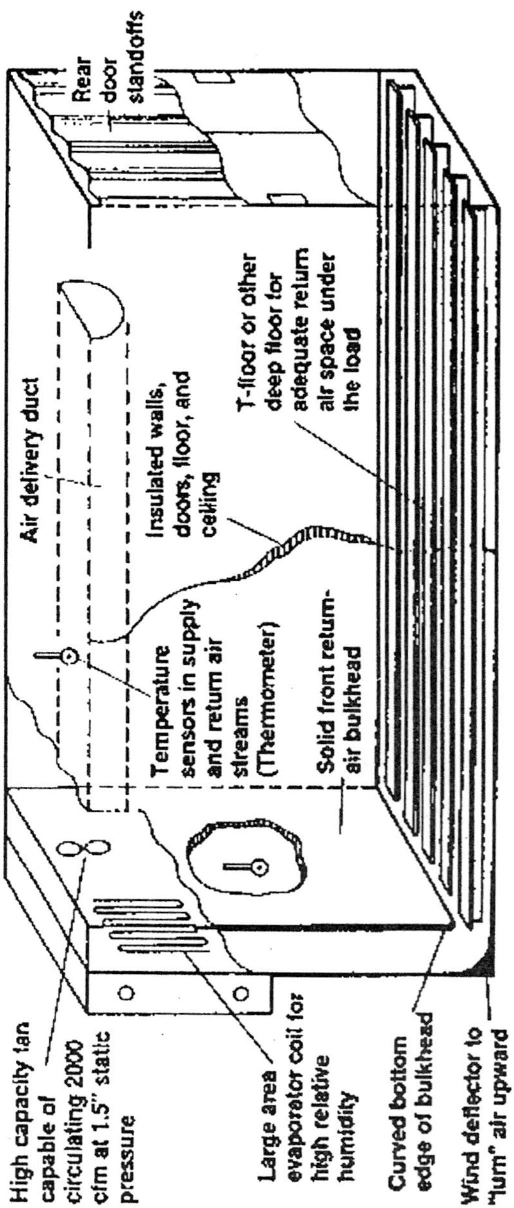

Figure 6. Features in a top-air delivery trailer

use of the vehicles in this way is a waste of a very expensive transport vehicle.

Stacking Patterns

Handstacked

In order to reduce the conduction of heat from outside the vehicle into the load of produce, stacking patterns should reduce the amount of contact between the produce and the floor and walls of the vehicle.

Illustrated below is pyramid stacking of bags inside a refrigerated trailer.

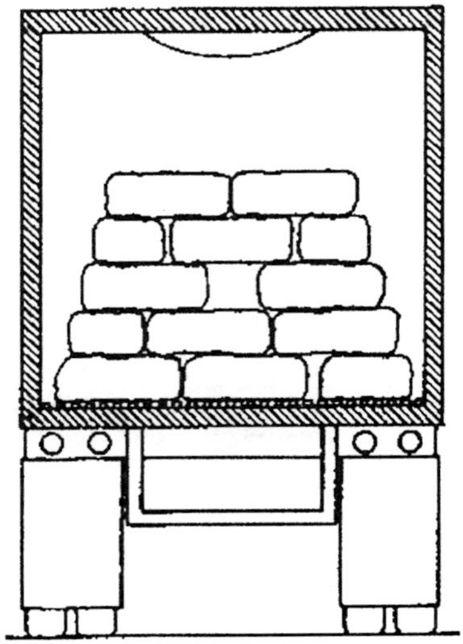

Figure 7. Pyramid stacking of bags inside a refrigerated trailer

Bushels of produce can be loaded into a refrigerated trailer using a pattern of alternately inverted layers that leave plenty of space between rows for air circulation.

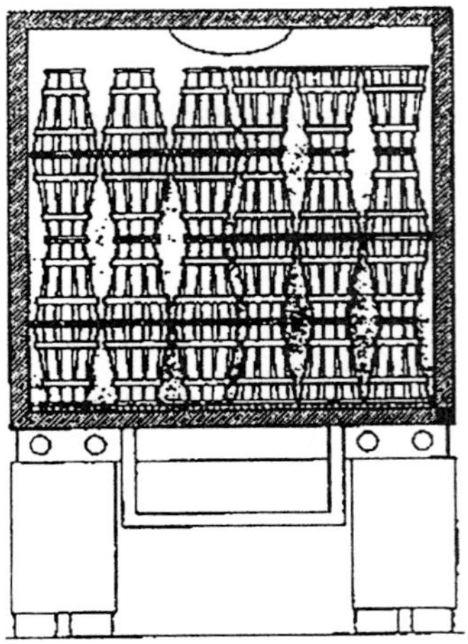

Figure 8. Pattern of alternately inverted layers

Produce transported in cartons should also be stacked so as to allow adequate air circulation throughout the load. The diagram below illustrates cross-wise offset loading of partial telescopic containers. On the floor of the truck, pallets or other supports should be used to keep the cartons out of direct contact with the floor.

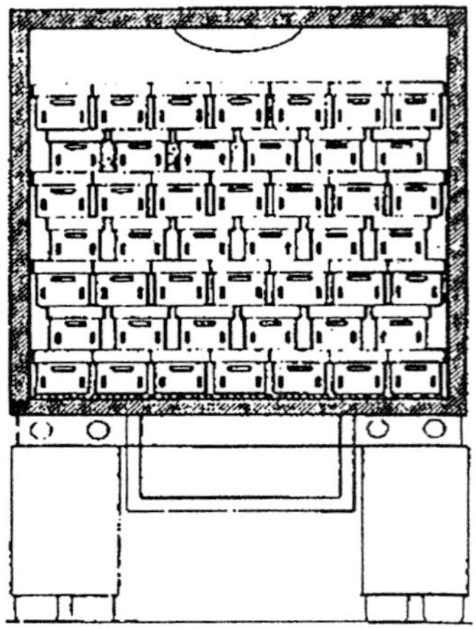

Figure 9. Cross-wise offset loading of partial telescopic containers

When cartons of various sizes must be loaded together, the larger, heavier containers should be placed on the bottom of the load. Parallel channels should be left for air to move through the length of the load.

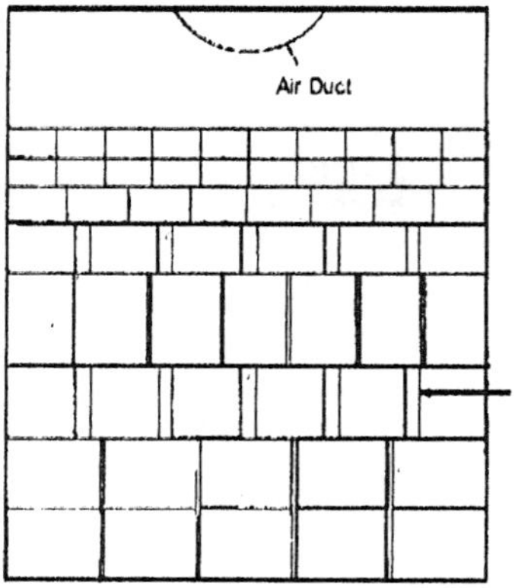

Figure 10. Parallel channels for air passage

Often the large containers used for cut flower packaging must be handstacked when loaded into a transport vehicle. The best loading pattern for cut flowers is known as the pigeon hole pattern, where boxes are stacked in alternating solid and open layers, and channels are left down both side walls.

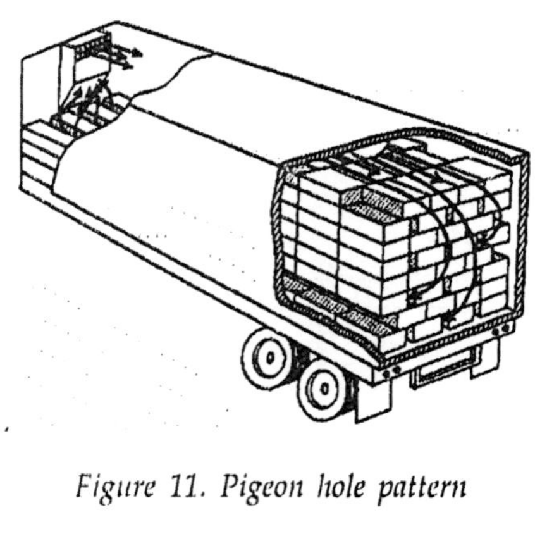

Figure 11. Pigeon hole pattern

This pattern provides channels for air circulation lengthwise through the load, and allows every box to be in direct contact with refrigerated air.

Pallet and Slip Sheet Loads

Containers should be loaded so that they are away from the side walls and the floor of the transport vehicle in order to minimise the conduction of heat from the outside environment. In the diagrams below, the numbers of cartons refer to how many cartons would be in contact with the walls and floor of the truck when fully loaded. Only the load on the bottom right is fully protected from heat transfer. The use of pallets keeps the cartons off the floor, while center-loading leaves an insulating air space between the pallet loads and the outside walls.

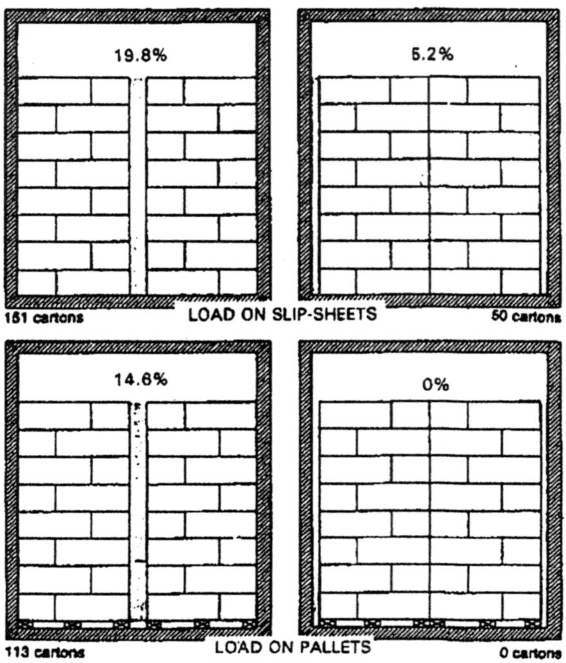

Figure 12. Pallet loads and slip sheets

Bracing the Load

There should always be a void between the last stack of produce and the back of the transport vehicle. The load should be braced to prevent shifting against the rear door during transit. If the load shifts, it can block air circulation, and fallen cartons can present great danger to workers who open the door at a destination market. A simple wooden brace can be constructed and installed to prevent damage during transport.

Sea Transport

The inter-island trade in fresh produce utilises a range of small vessels of wooden or steel construction which are independently owned and operated and focus almost exclusively on servicing the transport needs of the trade. The vessels all rely on engines for propulsion, but the wooden sloops regularly use their sails to stabilise and propell them. The wooden sloops are generally newer (ten years old or less) and carry only about 35 to 40 tonnes of cargo. The steel vessels by comparison are much older and most were bought cheaply and fairly recently second-hand from Europe where they were considered too old or expensive for continued service. Many of the steel vessels are 50 years or older and carry 60 to 100 or more tonnes of cargo.

Some of the larger steel vessels, such as the 'Stella SII' and the 'Louise Kingcraft', carry other types of cargo on a regular or opportunistic schedule between their regular fortnightly huckster route. Several of the small wooden sloops operating out of Dominica also double as fishing boats - fresh produce from Dominica is off-loaded in Antigua then the vessel will fish the waters in the Leewards and sell their catch at the weekend market in Antigua before returning to Dominica for more fresh produce.

All of the rest of the inter-island vessels concentrate solely on the fresh produce trade, but it should be noted that the return journey with dry goods and other consumables ranging from snack foods from Trinidad or cylinders of LPG gas and spare parts etc., is equally important to both the hucksters and traffickers in terms of cargo volume and probably more important in many cases in terms of profits.

Early in 1987, a Consultant Marine Engineer with the FAO Inter-Island Trade Project conducted a survey of, most of the vessels involved in the inter-island trade in fresh produce at that time. His conclusions may be summarised as follows:

— *General condition* - The condition of most of the vessels themselves was found to be reasonably good. Only the 'Bigdin' was found totally unsatisfactory, but then it was built in 1898 and should have been scrapped a long time ago. Many of the vessels regularly experienced engine breakdowns for lack of parts or occasionally lack of maintenance.

— *Fitness for fresh produce* - Not one single vessel of either steel or wooden construction was found to be appropriate for fresh produce transportation because of:

  — *Poor ventilation* - none of the vessels had adequate ventilation of the hold and most had no ventilation at all;

  — *No insulation of the engine compartment* - in most of the wooden vessels the engine compartment is not separated from the cargo hold and heat and fumes from the engine mixes with the cargo, often the engine sucks air from the cargo hold because there is no other air supply for engine operation;

— *No cargo liftinb equipment* - few of the vessels had equipment to assist with loading/ unloading of cargo and much damage was impacted on the produce from rough handling by the crew and stevedores;

— *Cargohold shape* - Only the larger steel vessels offered a practical shape for loading of crates and boxes and the wooden vessels shape meant that sacks and bags were crammed into the spaces between and under heavy wooden crates.

The FAO Consultant recommended various inexpensive and practical improvements for the vessels and these were presented to the captains and owners for their guidance. The FAO project offered to financially assist with fitting of one or two demonstration vessels with the prescribed improvements, but met with no response and to date none of the vessels has been improved at the owner's or captain's initiative. The improvements recommended were:

— *Improved ventilation* - by fitting "cowl-type" ventilators as shown in Figure 13 below.

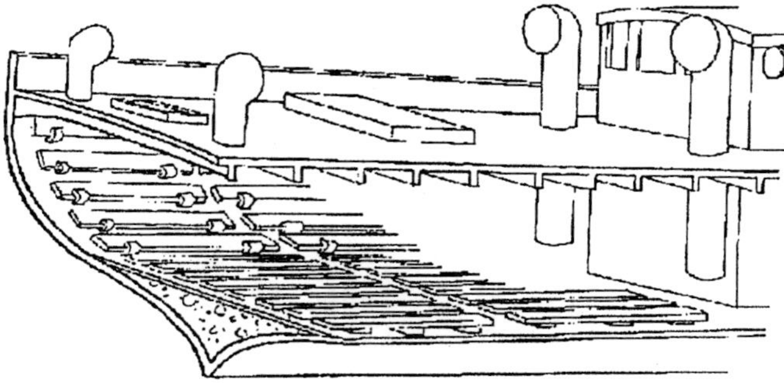

Figure 13. "Cowl-type" ventilators for inter-island vessels

Most of the vessels are not equipped with generators to provide power for electric fans and blowers, so natural airflow and some wind assistance can be used to direct air into the bottom of the cargo and engine holds separately and extract the warm air from the top of the holds. Cost is estimated at approximately US$2, 500 for each vessel assuming only one cargo hold.

— *Insulation of engine compartment* - by construction of a proper barrier between the engine and the cargo hold and fitting an insulating layer to prevent heat from the engine entering the cargo hold. The insulation could be prefabricated polystyrene panels, or steel coated polystyrene sandwich panels, or polyester foam insulation sprayed into place or onto spercially prepared plywood panels. All costs include wood, plywood and other materials as well as the insulation material itself.

— *Cargo lifting equipment* - intended only for the wooden sloops. Figure 14 below shows a suitable installation which can be fitted to the mast and yet would not interfere with normal sailing operations. At a later stage owners could fit a small diesel powered winch to further improve the design.

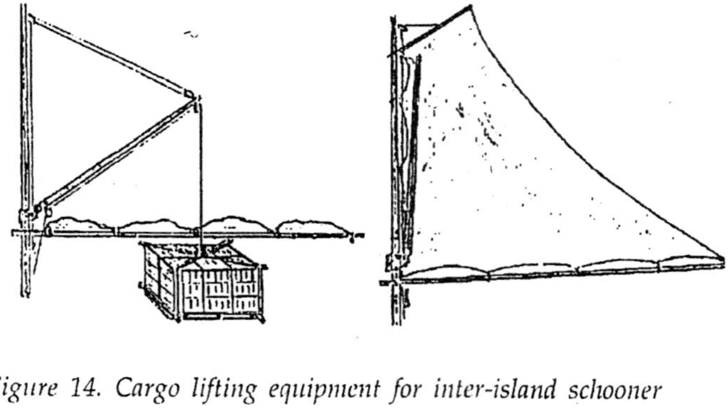

Figure 14. Cargo lifting equipment for inter-island schooner

Very few of the inter-island vessels carry any form of insurance because the owners claim the premiums are too high. The history of the inter-island trade is characterised by many sinking of wooden vessels with a complete loss of cargo and occasionally the lives of the hucksters accompanying their produce. Today, only a few of the steel vessels carry passengers and most of the traders fly to their destinations and meet the cargo there. The exceptions are still the very small open boats serving Marie Galante and Les Saintes from Dominica which still carry passengers and still experience losses of life and cargo.

Without insurance of their vessels, the owners cannot obtain loans for improvements to the vessels. However, the profits from the trade have been estimated on several occasions to be easily sufficient to support the cost of the above recommended improvements from cashflow alone and loans should not be necessary for any but the most poorly operated vessels. It would appear that the principal barrier to improving the vessels is the intransigence of the captains and owners content to reap their profits without the bother of improving the trade. Hucksters and traffickers generally lack sufficient organisation and cooperative spirit to force the owners into making improvements, and some owners are hucksters themselves.

Various reports, recommendations and proposals have been made by different aid organisations in the region to introduce an independent freight service based on modern and possibly purpose built vessels more suited to the transport of fresh produce. To-date, none of these suggested interventions has become a reality, and the inter-island trade will continue to suffer high postharvest losses of fresh produce because of inadequate transport service and conditions. The dilemma is such that some improvements can be made by introducing better forms of packaging but these will be limited until improvements are made in the vessels themselves. In addition, most of the vessels are really only suited to the use of sacks and bags because of the size and shape of the cargo hold.

Refrigerated Sea Transport

The perishability of fresh produce, allied with its tendency to heat up in confined spaces leading to rapid spoilage and decay, are all reasons why long distance unrefrigerated ship transportation is seldom used and never without high levels of spoilage. It is not likely that any mayor advances will be made in unrefrigerated shipping design to make the transport of fresh produce less risky. In most circumstances sea transportation is by reefer vessel and is largely used for export of fresh produce. Sea transportation, because of the journey times, is effectively a form of refrigerated storage and all the precautions necessary for storage are relevant here also.

Reefer vessels

Reefer vessels are totally refrigerated, have efficient air circulation systems and controllable rates of air-exchange. Loading is facilitated by side-hatches or by specialised loop-belt continuous conveyors, which transport the individual packages from the loading wharf up above the central hatches of the vessel and down into the holds (they are used in an identical fashion for off-loading). Reefer vessels are generally high capacity (4000 tones or more) and regularly carry fresh produce all over the world. The limiting factors are the journey times, which may exceed the storage-life of most produce, and the considerable amount of handling involved in loading and unloading. Palletisation of produce has reduced much of the handling but break-bulk loading of individual packages is still widely used.

Reefer vessel transportation usually involves the export of large volumes of fresh produce and necessitates the dedicated activities of full-time personnel employed by a large volume producer. For example in the Eastern Caribbean, the banana operations. Chartering, if to be successful means organising regular large volume supplies of produce over an extended period or charter vessels will not be available.

Reefer containers

Reefer containers are a specialised form of sea transportation which is rapidly growing in international popularity. Each container either has its own independent refrigeration system which can be powered electrically by the container vessel, or has special air ducts are one end which are lined up with conduits on the container vessel and refrigeration is thus provided entirely by the vessel's own system (this is know as the "ConAir" system).

Reefer containers are in standard sizes all of which are 8 x 8 ft. cross section, but which may be 10, 20, 30 or 40 ft. long. The most widely used sizes are the 40ft. followed by the 20ft. size. The container can be bought or leased but they can be very expensive and can vary greatly in quality and performance. Their principal advantages are:

— permit shared use of a reef container vessel by many producers of different commodities, provided they have access to containers and are exporting by the same route.

— greatly reduce handling damage since they are loaded at the packhouse and may not be unloaded until they reach a customer store in the country of export.

— temperature control is independently set and monitored.

— capable of rapid pre-cooling of produce from ambient tropical conditions.

However, their disadvantages are:

— very expensive to own or lease.

— large and heavy and require special lifting equipment.

— need to lease more containers than are in actual fuse because of delayed return and breakdown.

— not all countries have container handling facilities which may limit loading ports and discharge ports, although some vessels are self-loading and unloading.

Air Transportation

Air transportation is very expensive and usually can only be justified for high value export produce such as exotic tropical fruits and vegetables for the extra-regional markets. These markets are very sophisticated and demand top quality produce which is carefully packed in standardised fibre-board cartons and correctly labelled. Any produce not meeting these specifications, or of less than top quality, will either be rejected immediately or will be down-graded to a price level which gives a break-even price for the exporter or very often a loss on the consignment.

All air-freighted exports require a high degree of market research, planning, organisation and management. Constant communication with identified importers is vital to guage market trend, prices and fluctuations in demand, together with feedback on quality control.

Costs of air freight will vary according to distance carried and whether a charter or scheduled flight is used. If production and package costs are added it can very quickly be seen that profit margins are low. Some countries see export of these commodities less as a profit maker and more as an earner of valuable foreign exchange. Lack of planning and good management will result in poor quality, rejection and possible loss of income.

Airport Handling

Airports are designed to keep people away from cargo areas, but it is essential that personnel are given clearance for supervising the loading of produce onto the plane. The produce must be at the airport well before the time of flight departure. Delays are common and so facilities for holding the produce at the airport are necessary, and they must either be refrigerated or at the very least shaded. The loading supervisor or another person should have all necessary paperwork processed in good time and details of the consignment should be sent to the importer by telex or telefax.

Air Containers

Fresh produce may be sent by cargo carrier but more usually in the cargo hold of a passenger aircraft. The quantity that can be sent will vary according to aircraft and space available. Most aircraft servicing the Eastern Caribbean employ unit load containers (ULD), which fit the contours of the hold, as well as carrying goods packed loose. Others may use special thin aluminium pallets which are moved over rollers. These can be leased and a supply kept at the packhouse with the necessary securing nets. By palletising in this way at the packhouse and perhaps using refrigerated trucks the handling of the produce is greatly reduced and loading times for the aircraft become very rapid.

Temperature and Pressure

Cargo holds are frequently maintained at the same temperature and pressure as passenger areas, but this may not always be the case. Very often chill temperatures at high altitude can cause irreversible damage to the produce and flight conditions should always be checked with the airline. Of the aircraft currently servicing the Eastern Caribbean the situation tends to be confusing in that some holds may be temperature-controlled but at preset temperatures, and no choice is offered as to the temperature set. Other holds on the same aeroplane may have no temperature control and operate at ambient. Exporters must therefore select which produce must go in which hold well ahead of time and book that space with the airline if chilling of the produce is to be avoided.

Trans-shipment

Trans-shipment of fresh produce from one flight to another at an intermediate country airport is nearly always inviting disaster unless personnel are available to supervise. For example, air transport from Dominica and St. Vincent with only small air-fields must be trans-shipped out of the other islands airports which can accommodate large jets and have regular scheduled services to the extra-regional markets. The exporters must then rely upon the others to make sure that the trans-shipment occurs smoothly.

4

Preservation of Fruits and Vegetables

The magnitude of post harvest losses in fresh fruits and vegetables is estimated to be 25 - 80 % depending upon the commodity and the technological level of post harvest operations. This reflects the lack of knowledge by post harvest handlers of the biological and environmental factors involved in the deterioration or the absence of adequate post harvest technologies required preserving fresh quality.

Role of Water Activity (a_w)

The concept of a_w has been very useful in food preservation and on that basis many processes could be successfully adapted and new products designed. Water has been called the universal solvent as it is a requirement for growth, metabolism, and support of many chemical reactions occurring in food products. Free water in fruit or vegetables is the water available for chemical reactions, to support microbial growth, and to act as a transporting medium for compounds.

In the bound state, water is not available to participate in these reactions as it is bound by water soluble compounds such as sugar, salt, gums, etc. (osmotic binding), and by the surface effect of the substrate (matrix binding). These water-binding effects reduce the vapour pressure of the food substrate according to Raoult's Law. Comparing this vapour pressure with that of pure water (at the same temperature) results in a ratio called water activity (a_w). Pure water has an a_w of 1, one molal solution of sugar - 0.98, and one molal solution of sodium chloride - 0.9669.

A saturated solution of sodium chloride has a water activity of 0.755. This same NaCl solution in a closed container will develop an equilibrium relative humidity (ERH) in a head space of 75.5%. A relationship therefore exists between ERH and a_w since both are based on vapour pressure. The ERH of a food product is defined as the relative humidity of the air surrounding the food at which the product neither gains nor loses its natural moisture and is in equilibrium with the environment.

The definition of moisture conditions in which pathogenic or spoilage microorganisms cannot grow is of paramount importance to food preservation. It is well known that each microorganism has a critical a_w below which growth cannot occur. For instance, pathogenic microorganisms cannot grow at aw <0.86; yeasts and moulds are more tolerant and usually no growth occurs at a_w <0.62. The so-called intermediate moisture foods (IMF) have a_w values in the range of 0.65-0.90.

Enzyme-catalysed reactions can occur in foods with relatively low water contents. The apparent cessation of the reaction at low moisture cannot be due to the irreversible inactivation of the enzyme, because upon humidification to a higher water activity, hydrolysis resumes at a rate characteristic of the newly attained water activity.

The relationship between water content and water activity is complex. An increase in a_w is usually accompanied by an increase in water content, but in a non-linear fashion. This relationship between water activity and moisture content at a given temperature is called the moisture sorption isotherm. These curves are determined experimentally and constitute the fingerprint of a food system. Many methods and instruments are available for laboratory measurement of water activity in foods. Methods are based on the colligative properties of solutions. Water activity can be estimated by measuring the following:

- — Vapour pressure
- — Osmotic pressure
- — Freezing point depression of a liquid
- — Equilibrium relative humidity of a liquid or solid
- — Boiling point elevation
- — Dew point and wet bulb depression
- — Suction potential, or by using the isopiestic method
- — Bithermal equilibrium
- — Electric hygrometers
- — Hair hygrometers

Vapour Pressure

Water activity is expressed as the ratio of the partial pressure of water in a food to the vapour pressure of pure water with the same temperature as the food. Thus, measuring the vapour pressure of water in a food system is the most direct measure of a_w. The food sample measured is allowed to equilibrate, and measurement is taken by using a manometer or transducer device as depicted in Figure 1.

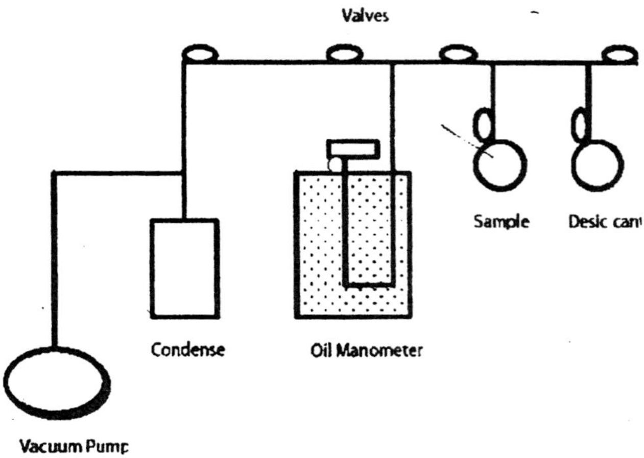

Figure 1. Vapour pressure manometer.

This method can be affected by sample size, equilibration time, temperature, and volume. This method is not suitable for biological materials with active respiration or materials containing large amounts of volatiles.

Depression and Elevation of Freezing Point

This method is accurate for liquids in the high water activity range but is not suitable for solid foods. The water activity can be estimated using the following two expressions:

Freezing point depression:

$$-\log a_w = 0.004207\ DT_f + 2.1\ E\text{-}6\ DT^2_f \qquad (1)$$

where DT_f is the depression in the freezing temperature of water.

Boiling point elevation:

$$-\log a_w = 0.01526\ DT_b - 4.862\ E\text{-}5\ DT^2_b \qquad (2)$$

where DT_b is the elevation in the boiling temperature of water.

Osmotic Pressure

Water activity can be related to the osmotic pressure (p) of a solution with the following equation:

$$p = RT/V_w \ln(a_w) \tag{3}$$

where V_w is the molar volume of water in solution, R the universal gas constant, and T the absolute temperature. Osmotic pressure is defined as the mechanical pressure needed to prevent a net flow of solvent across a semi-permeable membrane. For an ideal solution, Equation (3) can be redefined as:

$$p = RT/V_w \ln(X_w) \tag{4}$$

where X_w is the molar fraction of water in the solution. For non-ideal solutions, the osmotic pressure expression can be rewritten as:

$$p = RTfnmb(m_w V_w) \tag{5}$$

where n is the number of moles of ions formed from one mole of electrolyte, m_w and mb are the molar concentrations of water and the solute, respectively, and f the osmotic coefficient, defined as:

$$f = -m_w \ln(a_w)/nmb \tag{6}$$

Dew Point Hygrometer

Vapour pressure can be determined from the dew point of an air-water mixture. The temperature at which the dew point occurs is determined by observing condensation on a smooth, cool surface such as a mirror.

This temperature can be related to vapour pressure using a psychrometric chart.

Thermocouple Psychrometer

Water activity measurement is based on wet bulb temperature depression. A thermocouple is placed in the chamber where the sample is equilibrated. Water is then sprayed over the thermocouple before it is allowed to evaporate, causing a decrease in temperature. The drop in temperature is related to the rate of water evaporation from the surface of the thermocouple, which is a function of the relative humidity in equilibrium with the sample.

Isopiestic Method

The isopiestic method consists of equilibrating both a sample and a reference material in an evacuated desiccator until equilibrium is reached at 25°C. The moisture content of the reference material is then determined and the a_w obtained from the sorption isotherm. Since the sample was in equilibrium with the reference material, the a_w of both is the same.

Electric Hygrometers

Most hygrometers are electrical wires coated with hygroscopic salts or sulfonated polystyrene gel in which conductance or capacitance changes as the coating absorbs moisture from the sample. The major disadvantage of this type of hygrometer is the tendency of the hygroscopic salt to become contaminated with polar compounds, resulting in erroneous a_w determinations.

Hair Hygrometers

Hair hygrometers are based on the stretching of a fibre

when exposed to high water activity. They are less sensitive than other instruments at lower levels of activity (<0.03 a_w) and the principal disadvantage of these types of meters is the time delay in reaching equilibrium and the tendency to hysteresis. Today we find many brands of water activity meters in the market. Most of these meters are based on the relationship between ERH and the food system, but differ in their internal components and configuration of software used.

One of the water activity meters most used today is the AcquaLab Series 3 Model TE, developed by Decagon Devices, which is based on the chilled-mirror dew point method. This instrument is a temperature controlled water activity meter that allows placement of the sample in a temperature stable environment without the use of an external water bath. The temperature can be selected on the screen and is monitored and controlled with thermoelectric components. Most of the older generations of water activity instruments are based on a temperature-controlled environment. Therefore, a margin of error greater than 5% can be expected due to temperature variations. This equipment is highly recommended for measuring water activity in fruits and vegetables since it measures a wide range of water activity.

The major advantages of the chilled-mirror dew point method are accuracy, speed, ease of use and precision. The AquaLab's range is from 0.030 to 1.000a_w, with a resolution of ±0.001a_w and accuracy of ±0.003a_w. Measurement time is typically less than five minutes. Capacitance sensors have the advantage of being inexpensive, but are not usually as accurate or as fast as the chilled-mirror dew point method. Capacitive instruments measure over the entire water activity range 0 to 1.00 a_w, with a resolution of ±0.005a_w and accuracy of

±0.015a_w. Some commercial instruments can complete measurements in five minutes while other electronic capacitive sensors usually require 30 to 90 minutes to reach equilibrium relative humidity conditions.

Intermediate Moisture Foods (IMF)

Traditional intermediate moisture foods (IMF) can be regarded as one of the oldest foods preserved by man. The mixing of ingredients to achieve a given a_w, that allowed safe storage while maintaining enough water for palatability, was only done, however, on an empirical basis. The work done by food scientists approximately three decades ago, in the search for convenient stable products through removal of water, resulted in the so-called modern intermediate moisture foods. These foods rely heavily on the addition of humectants and preservatives to prevent or reduce the growth of microorganisms. Since then, this category of products has been subjected to continuous revision and discussion.

Definitions of IMF in terms of a_w values and moisture content vary within wide limits (0.6-0.90 a_w, 10-50% moisture), and the addition of preservatives provides the margin of safety against spoilage organisms tolerant to low a_w. Of the food poisoning bacteria, Staphylococcus aureus is one of the organisms of high concern since it has been reported to tolerate a_w as low as 0.83-0.86 under aerobic conditions. Many of the considerations on the significance of microorganisms in IMF are made in terms of a_w limits for growth. However, microbial control in IMF does not only depend on a_w but on pH, Eh, F and T values preservatives, competitive microflora, etc., which also exert an important effect on colonising flora.

The application of IMF technology has been very successful in preserving fruits and vegetables without refrigeration in most Latin American countries. For instance, the addition of high amounts of sugar to fruits during processing will create a protective layer against microbial contamination after the heat process. The sugar acts as a water activity depressor limiting the capability of bacteria to grow in food.

Food products formulated under this concept are stable at room temperature without thermal processing and can be generally eaten without rehydration. Some processed fruits and vegetables are considered IMF foods. These include cabbage, carrots, horseradish, potatoes, strawberries, etc.; their water activities at 30°C follow:

Foods	a_w
Cabbage	0.64
	0.75
Carrots	0.64
	0.75
Horseradish	0.75
Potatoes	0.75
	0.64
Strawberries	0.65
	0.75

Under these conditions, bacterial growth is inhibited but some moulds and yeast may grow at a_w greater than 0.70. In addition, chemical preservatives are generally used to inhibit the growth of moulds and yeasts in fruits and vegetables.

Advantages of IMF preservation are: Intermediate moisture foods have an a_w range of 0.65-0.90, and thus water activity is their primary hurdle to achieving microbial stability and safety. IMF foods are easy to prepare and store without refrigeration. They are energy efficient and relatively cheap. They are not readily subject to spoilage, even if packages have been damaged prior to opening, as with thermostabilised foods, because of low a_w. This is a plus for many developing countries, especially those in tropical climates with inadequate infrastructure for processing and storage, and offers marketing advantages for consumers all over the world.

Disadvantages of IMF preservation are: Some IMF foods contain high levels of additives (i.e., nitrites sulphites, humectants, etc.) that may cause health concerns and possible legal problems. High sugar content is also a concern because of the high calorific intake. Therefore, efforts are been made to improve the quality of such foods by decreasing sugar and salt addition, as well as by increasing the moisture content and a_w, but without sacrificing the microbial stability and safety of products if stored without refrigeration. This may be achieved by an intelligent application of hurdles.

Fruit products from intermediate moisture foods (IMF) appear to have potential markets. However, application of this technology to produce stable products at ambient temperature is limited by the high concentration of solutes required to reduce water activities to safe levels. This usually affects the sensory properties of the food.

Combined Methods for Preservation

Food preserved by combined methods (hurdles) remains stable and safe even without refrigeration, and is high in sensory and nutritive value due to the gentle process applied. Hurdle technology is the term often applied when foods are preserved by a combination of processes. The hurdle includes temperature, water activity, redox potential, modified atmosphere, preservatives, etc. The concept is that for a given food the bacteria should not be able to "jump over" all of the hurdles present, and so should be inhibited.

If several hurdles are used simultaneously, a gentle preservation could be applied, which nevertheless secures stable and safe foods of high sensory and nutritional properties. This is because different hurdles in a food often have a synergistic (enhancing) or additive effect. For instance, modified foods may be designed to require no refrigeration and thus save energy. On the other hand, preservatives (e.g., nitrite in meats) could be partially replaced by certain hurdles (such as water activity) in a food. Moreover, a hurdle could be used without affecting the integrity of food pieces (e.g., fruits) or in the application of high pressure for the preservation of other foods (e.g., juices). Hurdle technology is applicable both in large and small food industries.

In general, hurdle technology is now widely used for food design in making new products according to the needs of processors and consumers. For instance, if energy preservation is the goal, then energy consumption hurdles such as refrigeration can be replaced by hurdles (a_w, pH, or Eh) that do not require energy and still ensure a stable and safe product. The hurdle effect is an illustration of the fact that in most foods several factors (hurdles) contribute to stability and safety. This hurdle effect is of fundamental importance for the preservation of food, since the hurdles in a stable product control microbial spoilage and food poisoning as well as undesirable fermentation.

Increasing consumer demand for fresh quality products is turning processors to the so-called minimally processed products (MP), an attempt to combine freshness with convenience to the point that even the traditional whole, fresh fruit or vegetable is being packaged and marketed in ways formerly reserved for processed products. According to these authors, the widely accepted concept of MP refrigerated fruits involves the idea of living respiring tissues. Because MP refrigerated products can be raw, the cells of the vegetative tissue may be alive and respiring (as in fruits and vegetables), and biochemical reactions can take place that lead to rapid senescence and/or quality changes.

In these products, the primary spoilage mechanisms are microbial growth and physiological and biochemical changes, and in most cases, minimally processed foods are more perishable than the unprocessed raw materials from which they are made. The technology for shelf-stable high moisture fruit products (HMFP) is based on a combination of inhibiting factors to combat the deleterious effects of microorganisms in fruits, including additional factors to reduce major quality losses from reactions.

In order to select a combination of factors and levels, the type of microorganism and quality loss from reactions that might occur must be anticipated. Minimal processing may encompass pre-cut refrigerated fruits, peeled refrigerated whole fruits, sous vide dishes, which may include pre-heated vegetables and fruits, cloudy and clarified refrigerated juices, freshly squeezed juices, etc. All of these products have special packaging requirements coupled with refrigeration. These products, apart from special handling, preparation, and size reduction operations, might also require special distribution and utilisation operations such as Controlled atmosphere/Modified atmosphere/air flow rate/vacuum storage (O_2, CO_2, N_2, CO, C_2H_2, H_2O controls), computer controlled warehousing, retailing and food service, communications network, etc. HMFP fruits are less sophisticated than MPR fruits and should be priced lower when introduced commercially.

Careful selection of these processes should of course be made to find the appropriate methods suited to a particular rural or village situation.

Water Activity Reducing Agents

Glucose: Glucose is not a very good humectant due to the lower water holding capacity (WHC), which makes it difficult to obtain the isotherm curve at low a_w.

Fructose: Fructose has a higher water activity reduction capacity and therefore is more desirable as a humectant in stabilising food products.

Sucrose: Sucrose is one of the most studied sugars and is widely used in food systems, in the confectionary industry, both in the U.S. and Europe, but has a lower water activity reduction capacity compared to fructose.

The water reduction capacity of sugar and salts in their amorphous and anhydrous state at different a_w is presented in Table 1.

Table 1. Water activity reduction capacities of sugars and salts.

Moisture content (gH_2O/100 g Solids)								
	Anhydrous				Amorphous			
Sugars	a_w= 0.60	0.70	0.80	0.90	0.60	0.70	0.80	0.90
Sucrose	3.0	5.0	10.0	-	14.0	20.0	35.0	65.0
Glucose	1.0	3.5	7.5	12.5	1.0	3.5	8.0	22.0
Fructose	14.0	22.0	34.0	47.0	18.0	30.0	44.0	80.0
Lactose	0.01	0.01	0.05	0.10	4.5	4.7	4.7	-
Sorbitol (adsorption)	17.0	22.0	37.0	76.0	25.0	35.0	55.0	110.0
Corn syrup	-	-	-	-	14.0	20.0	30.0	54.0

Contd....

Contd....

Salts								
NaCl (adsorption)	0.1	0.1	130.0	585.0	-	-	-	-
NaCl (desorption)	-	_	385.0	590.0	-	-	-	-
KCl (adsorption)	0.1	0.1	0.1	0.1	-	-	-	-
KCl (desorption)	-	-	0.1	580.0	-	-	-	-

Other humectants: Based solely on the water activity reduction capacity (Table 1), sorbitol and fructose are the most desirable humectants. Sucrose has the third best reduction capacity and lactose the poorest. The amorphous form absorbs more water at specific a_w than the corresponding crystalline form. As seen in Table 1, NaCl and KCl salts appear to be superior humectants at a high range of a_w. The increased a_w lowering ability exhibited by the salts may be explained by the smaller molecular weight, increasing the ability to bind or structure more water.

Other sugars used as humectants in food stability include lactose and sorbitol. The amorphous form absorbs more water at specific a_w than the crystalline form. Polyols are better humectants than sugars because of their greater water activity reduction capacity and are less hygroscopic than sugars. The most widely used polyols as humectants in foods are 1,3- butyleneglycol, propylene glycol, glycerol, and polyethylene glycol 400.

pH Reducing Agents

Organic acids: Organic acids, whether naturally present in foods due to fermentation or intentionally added during processing, have been used for many years in food preservation. Some organic acids behave primarily as fungicides or fungistats, while others tend to be more effective at inhibiting bacterial growth. The mode of action of organic acids is related to the pH reduction of the substrate, acidification of internal components of cell membranes by ionisation of the undissociated acid molecule, or disruption of substrate transport by alteration of cell membrane permeability.

The undissociated portion of the acid molecule is primarily responsible for antimicrobial activity; therefore, effectiveness depends upon the dissociation constants (pKa) of the acid. Organic acids are generally more effective at low pH and high dissociation constants. The most commonly used organic acids in food preservation include: citric, succinic, malic, tartaric, benzoic, lactic, and propionic acids.

Citric acid is present in citrus fruits. It has been demonstrated that citric acid is more effective than acetic and lactic acids for inhibiting growth of thermophilic bacteria. Also, combinations of citric and ascorbic acids inhibit growth and toxin production of C. botulinum type B in vacuum-packed cooked potatoes.

Malic acid is widely found in fruits and vegetables. It inhibits the growth of yeasts and some bacteria due to a decrease in pH.

Tartaric acid is present in fruits such as grapes and pineapples. The antimicrobial activity of this acid is attributed to pH reduction.

Benzoic acid is the oldest and most commonly used preservative. It occurs naturally in cranberries, raspberries, plums, prunes, cinnamon, and cloves. As an additive, sodium salt in benzoic acid is suitable for foods and beverages with pH below 4.5. Benzoic acid is primarily used as an antifungal agent in fruit-based and fruit beverages, fruit products, bakery products, and margarine.

Lactic acid is not naturally present in foods; it is formed during fermentation of foods such as sauerkraut, pickles, olives, and some meats and cheeses by lactic acid bacteria. It has been reported that lactic acid inhibits the growth of spore forming bacteria at pH 5.0 but does not affect the growth of yeast and moulds.

Propionic acid occurs in foods by natural processing. It is found in Swiss cheese at concentrations up to 1%, produced by Propionicbacterium shermanii. The antimicrobial activity of propionic acid is primarily against moulds and bacteria.

Inorganic acids: Inorganic acids include hydrochloric, sulphuric, and phosphoric, the latter being the principal acid used in fruit and vegetable processing). They are mainly used as buffering agents, neutralisers, and cleaners.

Fermentation by-products: Fermentation by-products are formed during fermentation of fruits and vegetables, as in sauerkraut processing, pickling, and wine making. One by-product, lactic acid, is formed during fermentation of cabbage or cucumbers. This acid decreases the pH of fruits and vegetables, producing the characteristic flavour of sauerkraut, and acts as a controller of pathogens that may develop in the final fermented product.

Browning Inhibitors

Sulphites, bisulphites, and metabisulphites: Sodium bisulphite is a potential browning inhibitor in fruit and vegetable products (e.g., peeled potatoes and apples). This preservative when used in food production can delay or prevent undesirable changes in the colour, flavour, and texture of fresh fruits and vegetables, potatoes, drinks, wine, etc. Potassium bisulphite is used in a similar way to sodium bisulphite, and is used in the food industry to prevent browning reactions in fruit and vegetable products. Sulphites, bisulphites, and metabisulphites of both sodium and potassium together with gaseous sulphur dioxides are all chemically equivalent.

Sulphite levels in processed foods are expressed as SO_2 equivalents, and range from zero to about 3000 ppm in dry weight. Dehydrated, light coloured fruits (e.g., apples, apricots, bleached raisins, pears, and peaches) contain the greatest amounts in this range. Dehydrated vegetables and prepared soup mixes range from a few hundred to about 2000 ppm; instant potatoes contain approximately 400 ppm. The dose for wine is about 100-400 ppm and for beer about 2-8 ppm. The maximum legal sulphite level in wines permitted by the Food and Drug Administration (FDA) is 300 ppm. In the U.S. most wines have a sulphite level of 100 ppm.

Sulphites are highly effective in controlling browning in fruits and vegetables, but are subject to regulatory restrictions because of adverse effects on health. Sulphites inhibit non-enzymatic browning by reacting with carbonyl intermediates, thereby preventing further reaction. Sulphite levels in foods vary widely depending on the application. Residual levels never exceed several hundred per million but could reach 100 ppm in some fruits and vegetables. The maximum sulphur dioxide levels in fruit juices, dehydrated potatoes, and dried fruits permitted by the FDA are 300, 500, and 2000 ppm, respectively.

Additives

Potassium sorbate: Potassium sorbate is a white crystalline powder that has greater solubility in water than sorbic acid, which may be used accordingly in making concentrates for dipping, spraying, or metering fruit and vegetable products. It has antimycotic actions similar to sorbic acid, but usually 25% more potassium sorbate must be used than sorbic acid to secure the same protection. The common salt of potassium sorbate was developed because of its high solubility in water, which is 58.2% at 20°C.

In water, the salt hydrolysis yielded is the active form. Stock solutions of potassium sorbate in water can be concentrated up to 50%, which can be mixed with liquid food products or diluted dips and sprays. Sorbates are effective in retarding the growth of many food spoilage organisms. Sorbates have many uses because of their milder taste, greater effectiveness, and broader pH range (up to 6.5), when compared to either benzoate or propianate. Thus, in foods with very low pH, sorbate levels as low as 200 ppm may give more than adequate protection. The solubility of potassium sorbate is 139 g/ 100 mL at 20°C; it can be applied in beverages, syrups, fruit juices, wines, jellies, jams, salads, pickles, etc.

Sodium benzoate: The use of sodium benzoate as a food preservative has been limited to products that are acid in nature. Therefore, it is mainly used as an antimycotic agent (most yeasts and moulds are inhibited by 0.05-0.1%). The benzoates and parabenzoates have been used primarily in fruit juices, chocolate syrup, candied fruit peel, pie fillings, pickled vegetables, relishes, horseradish, and cheeses. Sodium benzoate is more effective in food systems where the pH is as low as 4.0 or below.

Other additives: Other naturally antimicrobial compounds found in fruits and vegetables include:

- Vanillin (4-hydroxi-3-methoxybenzaldehyde) is found primarily in vanilla beans and in the fruit of orchids (Vanilla planifola, Vanilla pompona, or Vanilla tahitensis). Vanillin is most active against moulds and non-lactic acid gram-positive bacteria. The effectiveness of vanillin against certain moulds such as A. flavus, A. niger, A. ochraceus, or A. parasiticus has been demonstrated in laboratory media, as well as its effectiveness against yeasts such as Saccharomyces cerevisiae, Pichia membranaefaciens, Zygosaccharomyces bailii, Z. rouxii and Debaryomyces hansenii.
- Allicin is an antimicrobial present in the juice vapour of garlic. This compound is effective in inhibiting the growth of certain pathogenic bacteria such as B. cereus, C. botulinum, E. coli, Salmonellae, Shigellae, S. aureus, A. flavus, Rhodotorula, and Saccharomyces.
- Cinnamon and eugenol are reported to have an inhibitory effect on the spores of Bacillus anthracis. Also, cinnamon was found to inhibit the growth of the aflatoxin of A. parasiticus. Aqueous clove infusions of 0.1 to 1.0% and 0.06% eugenol were reported to inhibit the growth of germinated spores of B. subtilis in nutrient agar.
- Oregano, thyme, and rosemary have been found to have inhibitory activity against certain bacteria and moulds due to the presence of antimicrobial compounds in their essential oils (e.g., terpenes, carvacol, and thymol).

Thermal Treatment

The main function of heat in food processing is to inactivate pathogenic and spoilage organisms, as well as enzyme inactivation to preserve foods and extend shelf life. Other advantages of heat processing include the destruction of anti-nutritional components of foods (e.g., trypsin inhibitors in legumes), improving the digestibility of proteins, gelatinisation of starches, and the release of niacin. Higher temperatures for shorter periods achieved the same shelf life extension as food treated at lower temperatures and longer periods, and allowed retention of sensory and nutritional properties.

Hot water: Hot water plays an important role in the sanitation of food products before processing. Some food products are treated with hot water to eliminate insects, and to inactivate microorganisms and enzymes. Foods are retained in a water blancher at 70-100°C for a specific time and then removed to a dewatering and cooling system.

Steam: Steam is a more effective means than hot water for blanching foods such as fruits and vegetables. This method is especially suitable for foods with large areas of cut surfaces. It retains more soluble compounds and requires smaller volumes of waste for disposal than those from water blanchers. This is particularly so if air-, rather than water-cooling is used. Furthermore, steam blanchers are easier to clean and sterilise.

Temperatures ranging from 10 to 15°C above the optimum temperature for growth will destroy vegetative cells of bacteria, yeasts, and moulds. Most vegetative cells, as well as viruses, are destroyed when subjected to temperatures of 60 to 80°C for an appropriate time. Somewhat higher temperatures may be needed for thermophilic or thermoduric microorganisms. All vegetative cells are killed in 10 min at 100°C and many spores are destroyed in 30 min at 100°C. Some spores, however, will resist heating at 100°C for several hours.

Freezing

Freezing is a quick and convenient way to preserve fruits and vegetables at home. It is a popular method of home food preservation. Home frozen fruits and vegetables of high quality and maximum nutritional value can be produced if the directions below are followed. These directions are based on:

- the chemical and physical reactions which take place during the freezing process;
- scientific knowledge of the effect of freezing on the tissues of fruits and vegetables; and
- food microbiology.

Fresh fruits and vegetables, when harvested, continue to undergo chemical changes which can cause spoilage and deterioration of the product. This is why these products should be frozen as soon after harvest as possible and at their peak degree of ripeness. Fresh produce contains chemical compounds called enzymes which cause the loss of colour, loss of nutrients, flavour changes, and colour changes in frozen fruits and vegetables. These enzymes must be inactivated to prevent such reactions from taking place.

Enzymes in vegetables are inactivated by the blanching process. Blanching is the exposure of the vegetables to boiling water or steam for a brief period of time. The vegetable must then be rapidly cooled in ice water to prevent it from cooking. Contrary to statements in some publications on home freezing, in most cases blanching is absolutely essential for producing quality frozen vegetables. Blanching also helps to destroy microorganisms on the surface of the vegetable and to make some vegetables, such as broccoli and spinach, more compact.

The major problem associated with enzymes in fruits is the development of brown colours and loss of vitamin C. Because fruits are usually served raw, they are not blanched like vegetables. Instead, enzymes in frozen fruit are controlled by using chemical compounds which interfere with deteriorative chemical reactions. The most common control chemical is ascorbic acid (vitamin C). Ascorbic acid may be used in its pure form or in commercial mixtures with sugars.

Some directions for freezing fruits also include temporary measures to control enzyme-activated browning. Such temporary measures include soaking the fruit in dilute vinegar solutions or coating the fruit with sugar and lemon juice. However, these latter methods do not prevent browning as effectively as treatment with ascorbic acid.

Another group of chemical changes that can take place in frozen products is the development of rancid oxidative flavours through contact of the frozen product with air. This problem can be controlled by using a wrapping material which does not permit air to pass into the product. It is also advisable to remove as much air as possible from the freezer bag or container to reduce the amount of air in contact with the product.

Textural Changes

Water makes up over 90 percent of the weight of most fruits and vegetables. This water and other chemical substances are held within the fairly rigid cell walls which give support structure, and texture to the fruit or vegetable. Freezing fruits and vegetables actually consists of freezing the water contained in the plant cells. When the water freezes, it expands and the ice crystals cause the cell walls to rupture.

Consequently, the texture of the produce, when thawed, will be much softer than it was when raw. This textural difference is especially noticeable in products which are usually consumed raw. For example, when a frozen tomato is thawed, it becomes mushy and watery. This explains why celery, lettuce, and tomatoes are not usually frozen and is the reason for the suggestion that frozen fruits, usually consumed raw, be served before they have completely thawed. In the partially thawed state, the effect of freezing on the fruit tissue is less noticeable. Textural changes due to freezing are not as apparent in products which are cooked before eating because cooking also softens cell walls. These changes are also less noticeable in high starch vegetables, such as peas, corn, and lima beans.

The extent of cell wall rupture can be controlled by freezing produce as quickly as possible. In rapid freezing, a large number of small ice crystals are formed. These small ice crystals produce less cell wall rupture than slow freezing which produces only a few large ice crystals. This is why some home freezer manuals recommend that the temperature of the freezer be set at the coldest setting several hours before foods will be placed in the freezer.

Some freezer manuals tell the location of the coldest shelves in the freezer and suggest placing unfrozen products on these shelves. All freezer manuals give guidelines for the maximum number of cubic feet of unfrozen product which can be frozen at one time. This is usually 2 to 3 pounds of vegetable to each cubic foot of freezer space per 24 hours. Overloading the freezer with unfrozen products will result in a long, slow freeze and a poor quality product.

Temperature Fluctuation

To maintain top quality, frozen fruits and vegetables

should be stored at 0° F or lower. This temperature is attainable in separate freezer units and in some combination refrigerator-freezers. A freezer thermometer can help you determine the actual temperature of your freezer. Storing frozen foods at temperatures higher than 0° F increases the rate at which deteriorative reactions can take place and can shorten the shelf life of frozen foods.

Do not attempt to save energy in your home by raising the temperature of frozen food storage above 0° F. Fluctuating temperatures in the freezer can cause the migration of water vapour from the product to the surface of the container. This defect is sometimes found in commercially frozen foods which have been improperly handled.

Loss of Moisture

Moisture loss, or ice crystals evaporating from the surface area of a product, produces freezer burn—a grainy, brownish spot where the tissues become dry and tough. This surface freeze-dried area is very likely to develop off flavours. Packaging in heavyweight, moistureproof wrap will prevent freezer burn.

Growth of Microbes

The freezing process does not actually destroy the microorganisms which may be present on fruits and vegetables. While blanching destroys some microorganisms and there is a gradual decline in the number of these microorganisms during freezer storage, sufficient populations are still present to multiply in numbers and cause spoilage of the product when it thaws. For this reason it is necessary to carefully inspect any frozen products which have accidentally thawed by the freezer going off or the freezer door being left open.

Nutrient Value

Freezing, when properly done, is the method of food preservation which may potentially preserve the greatest quantity of nutrients. To maintain top nutritional quality in frozen fruits and vegetables, it is essential to follow directions contained in this leaflet for pretreatment of the vegetables, to store the frozen product at 0°F and to use it within suggested storage times.

Storage Times

Most frozen fruits maintain high quality for 8 to 12 months. Unsweetened fruits lose quality faster than those packed in sugar or sugar syrups.

Most vegetables will maintain high quality for 12 to 18 months at 0° F or lower. However, it is a good idea to plan to use your home frozen vegetables before the next year crop is ready for freezing.

Longer storage of fruits and vegetables than those recommended above will not make the food unfit for use, but will decrease its quality.

Quality of Freezer Containers

You must use good quality freezer containers to maintain the quality of frozen fruits and vegetables. A high quality wrap should be both moisture and vapour proof so that moisture can be kept in the product and air kept away from it. Many moisture- and vapour-resistant wraps, such as heavyweight aluminum foil, plastic coated freezer paper, saran, and other plastic films, are effective at excluding oxygen. They should be strong, pliable, and adhere to the shape of the food item. These can be sealed easily with heat or freezer tape.

Be sure to use only tape that is designated for the freezer because other household tapes lose adhesive quality in the extremely cold freezer temperatures. These wraps are not as convenient for fruits and vegetables as plastic bags or rigid freezer containers. Plastic film bags made especially for freezing are readily available. They seal with twist and tie tops. Collapsible cardboard freezer boxes are frequently used as an outer covering for plastic bags to protect them against tearing, and for easy stacking in the freezer. Plastic sandwich bags and bread wrappers are not suitable for freezing.

"Freeze-and-cook" bags withstand temperatures from below 0° F to above the boiling point and are suitable for both freezing and cooking the product. These come in 1½ pint and quart sizes and also as large rolls of plastic so that they can be made the size desired. A heat sealer is necessary for closing these bags. These products are more expensive but allow greater convenience.

Packing Methods

There are three ways to pack fruits for freezing: sugar pack, syrup pack, and unsweetened pack. Although some fruits may be packed without sweeteners, the flavour of many fruits is retained better with the use of sugar. Gooseberries, currants, cranberries, blueberries, and rhubarb give good quality packs without or with sugar.

To freeze fruits using sugar pack, sprinkle the required amount of sugar over the fruit. Gently stir until the pieces are coated with sugar and juice. To make sugar syrup, dissolve the needed amount of sugar in cold water. Stir the mixture and let stand until the solution is clear.

There are two basic methods for packing vegetables for freezing, the tray pack and the dry pack.

Dry pack: This is the method used to describe the packing of blanched and drained vegetables into containers or freezer bags. Pack the vegetables tightly to cut down on the amount of air in the container. If the vegetables are packed in freezer bags, press air out of the unfilled part of the bag. When packing broccoli, alternate the heads and stems.

Tray Pack: This is the method of freezing individual pieces of blanched and drained vegetables on a tray or shallow pan, then packing the frozen pieces into a freezer bag or container. This method produces a product similar to commercially frozen plastic bags of individual vegetable pieces and is particularly good for peas, corn, and beans.

In this method it is most important to pack the individually frozen pieces into a bag or container as soon as they are frozen.

Necessary equipment for processing vegetables are:

- a large kettle (minimum capacity of 2 gallons)
- a colander, wire basket, or net bag for blanching
- large pans for cooling
- ice cubes or ice blocks for cooling
- knives
- plastic freezer bags or other containers
- a timer or a clock with a second hand
- hot pads

Choose vegetables for freezing that are at their peak of flavour and texture. If possible, harvest the vegetables in the cool part of the morning and process as quickly as possible. If the freezing process is delayed, immerse the vegetables in very cold water or refrigerate in shallow trays to preserve quality and nutrients. Carefully follow the blanching instructions in the included table for each vegetable. Count the blanching time from when the vegetable is immersed in the vigorously boiling water. The quality of water used to blanch the vegetables can have an effect on the texture of certain vegetables. Very hard water can cause the toughening of vegetables such as green beans. If you have problems with excessively tough green beans, check into the level of hardness in your water supply.

Boiling Water Blanching

- Use 1 gallon water for each pound of vegetable except for leafy greens, which need 2 gallons per pound.
- Bring water to rolling boil.
- Immerse wire basket or blanching basket mesh bag containing vegetable.
- Cover kettle and boil at top heat the required length of time. Begin counting time as soon as you place the vegetable in water. You may use the same blanching water 2 or 3 times. Keep it at required level. Change the water if it becomes cloudy.
- Cool immediately in ice water for same time used for blanching. Keep chilling water ice cold.
- Drain the vegetables thoroughly. Extra water will form too many ice crystals.
- Pack using dry or tray pack method.
- Freeze.

Steam Blanching

- Put 1 inch of water in kettle, bring to a rolling boil.

- Suspend a thin layer of vegetable in wire basket or loose cheesecloth over rapidly boiling water.
- Cover and steam blanch vegetable required amount of time as listed on table.
- Complete as for boiling water blanching.

Oven Blanching

Some directions are available for microwave blanching of vegetables. Incomplete inactivation of enzymes may occur due to the variability of microwave heating resulting in a shorter storage time in the freezer. Lack of standardisation of power levels among various microwave ovens make it impossible to publish a blanching timetable that can be used with all microwave ovens. Follow the instructions in the manufacturer's microwave cookbook. Vegetables blanched in the microwave should be chilled in ice water and processed as regular frozen vegetables.

To Use Home Frozen Produce

Fruits: Thaw fruit at room temperature in its original package to preserve quality and nutritive value. If faster defrosting is required, submerge (if watertight) in cool or lukewarm water or follow microwave defrosting instructions. Serve as soon as defrosted, preferably while a few ice crystals remain.

Vegetables: All vegetables may be cooked from the frozen state except corn-on-the-cob, which should be partially defrosted. Cook frozen vegetables in a small amount of salted water (about ½ cup or less). Cook only until tender—about half as long as if the same vegetable were fresh. You can use a pressure saucepan for cooking frozen vegetables. Follow manufacturer's directions for cooking time. A pack should be thawed enough to break it up before pressure cooking.

Wax Coating

In recent years, much attention has been paid to explore the potential of surface coating to maintain the quality of harvested fresh produce and to reduce the volume of disposable non-biodegradable packaging materials. Many fruits develop a waxy coat on their epidermis as they mature on the plant (apples, bananas, mangoes, and grapes) they develop a waxy coating on the epidermis, which is either felt or seen as a powdery bloom. Wax usually develops when the fruit has attained 2/3 of its growth. But much extraneous matter and spray residues.

The natural waxy coat is not adequate to offer protection against water loss and high respiration rate. Thus the approach towards prolonging the life of fruit / vegetable involves restricting the rate of respiration and preventing moisture loss. So as to maintain the vital food elements in as near the same quality as in the freshly picked fruit or vegetable.

The growing importance of fruits and vegetables in commerce has led to development in various protective wax coatings with and without fungicides, bactericides, growth regulators, etc., wax coating of fresh fruit and vegetable is quite old and it has great potential in the storage and transportation of fresh produce. Surface coatings have been used extensively on bulky organs to modify internal atmosphere composition and thereby delay ripening, reduce the water loss, and it improves the finish of the skin.

Post harvest losses can be reduced to some extent by, selection of varieties, careful handling, cleaning and sorting, control of ripening process, control of spoilage, packaging, use edible films, use of fungicide, use of ethylene absorbent, use of sprout inhibitors, use of growth regulators, irradiation, packaging home operations, dumping, washing and waxing.

Waxing of fruits and vegetables is done according to the recommendations. Food grade waxes are used to replace some of the natural; waxes removed in washing and cleaning operations, and this helps to reduce the water loss during handling and marketing. If produce is waxed, the wax coating is allowed to dry thoroughly before further handling. Waxing of the produce is an old age art, which was started in the beginning of the 19th century. Waxing was used as a preservation technique for fruits since 1900. Wax has been used since pre historic times.

In 4200 BC, the ancient Egyptians kept bees and used the wax from honeycombs to make models and to preserve their dead and in encaustic painting. The Romans used in pottery making textile printing, as early as 12th century, the Chinese were using waxing processes on fruits, but it was not until 1922 that a waxing process was introduces for wide spread use on fruits and vegetables.

Wax is an ester a long chain aliphatic acid chain aliphatic alcohol. Waxes are the esters formed from a fatty acid and a high molecular weight alcohol. Principle involved during storage: The produce continues respiring and uses up all the O_2 with in the fruit which is not being replaced as quickly as before waxing because of the coating and produces CO_2, which accumulates within the fruit because it cannot escape as easily through the coating. Eventually the fruit will switch to anaerobic respiration that does not require O_2. An aerobic respiration produces much more CO_2, acetaldehyde and ethanol. The acetaldehyde and ethanol give off - flavour to the product, which are detrimental to the perceived quality. The natural barrier of the fruit and vegetable, the type of wax and amount if was applied will influence the extent to which the internal atmosphere (O_2 and CO_2) are modified and the level of reduction in weight loss.

All fruits and vegetable are living organisms and as such require intake of O_2 and release CO_2 for respiration (aerobic). They also release vapour top the atmosphere through diffusion. Both these processes create a loss in weight of the produce. All fruit and vegetables are covered naturally in a cuticle, which is barrier to gas exchange. The cuticle is very impermeable to O_2, slightly more permeable to CO_2 and several magnitudes more permeable to water vapour.

Unwaxed fruit or vegetables have pores in their surface through which most the diffusion of O_2 and CO_2 offers. Water vapour can move through the pores, cuticle and micro-cracks in the cuticle. During the process of waxing a tightly adhering thin film of the coating substance is applied to the surface f the fruit. The surface coating is often relatively impermeable to O_2 and CO_2 and water.

Human became more health conscious. This gave new awareness of fresh fruits and vegetables, and resulted in rapidly increasing demand. Consumption of fresh produce, to help the consumers to obtain possible product, Fruits and vegetables are living organisms, which contain 80 - 90 % water by weight. If they are left without cuticle, the water quickly begins to evaporate, resulting in poor product shelf life to increase freshness and appearance. The fruit and vegetable growers should concern about consumer safety. There are few active ingredients which when combined with water, wetting and drying agents and form the compounds used a protective coating. The coating may contain one or more of the following. Carnauba wax, shellac wax, paraffin wax, candelilla wax, bees wax.

Types of Waxes

Animal waxes

1. Bees wax.
2. Spermaceti wax
3. Shellac wax
4. Chinese insect wax

Vegetable waxes

1. Carnauba wax
2. Candelilla wax
3. Sugarcane wax
4. Palm wax
5. Esparto wax
6. Japan wax
7. Oricury wax
8. Waxol 0.12

Mineral and synthetic wax

1. Ozocerite
2. Montan wax
3. Synthetic wax

Foodstuffs that are waxed are:

Fruits:

Apple Avocadoes Bell pepper Lemon Grapes

Banana Melons Oranges Lime Passion fruit

Peaches Pine apples

Vegetables:

Cucumber Tomato Sweet potato Melon

Methods of Waxing

Principle methods of waxing of fruits and vegetables are:

1. *Liquid Paraffin wax method*: In this method fruits and vegetables are dipped in hot paraffin. Some times resins are added. The main disadvantage of this method is too much of coating material is used.
2. *Slab wax method*: In this case the wax is pressed against rapidly revolving brushes. But the efficiency is very less.
3. *Spray method*: Spraying of melted wax on the fruit, which is subsequently brushed mechanically until a film of desired thickness is obtained. The wax is dissolved in a suitable solvent. This depends on,
   — The pressure employed
   — Volume of wax used
   — Wax temperature
   — Distance of fruit from the spray
   — Number of spray nozzles.
4. *Dipping or cold wax method*: Fruits and vegetables are washed and then without being dried are dipped into a wax emulsion of proper concentration. They are dried before packing. Purified wax is odour less, tasteless and nontoxic and it can be heat- sealed.

Properties of Wax

— Kneadable at 20° C.

— Easily emulsifiable.

— Should not impart undesirable odour.

— Should be economical.

— Efficient drying performance.

— Non-sticky or tacky.

— Should never interfere with the quality of fresh fruit/vegetable.

— Melts above 40° C without decomposition.

— Has relatively low viscosity.

— Capable of being polished by slight pressure.

— Translucent to opaque form but not like glass.

The properties of wax depend primarily on molecular structure rather than molecular size and chemical constitution.

Functions of Waxing

— To reduce the weight loss.

— To replace the natural wax.

— To increase the appearance

— To increase the freshness

— To decrease the rate of transpiration

— To reduce the shrinkage losses.

— To inhibit mold growth.

— To prevent other physical damage and disease.

During harvesting, transport, washing, grading, it can undergo some scratching or abrasion, which removes natural layer or was and also scars the protective layer of skin.

Lipids and Resins

Lipids and resin are added to coating formulations to reduce gas exchange, but mainly to impart hydrophobicity (to reduce water loss) and gloss appearance. Fruits coated with resins will develop whitening of the skin due to condensation that develops when they are brought from cold storage to ambient conditions.

Composite and Bilayer

Several composite and bilayer films have been investigated with goal of combining the desirable properties of different material to improve permeability characteristics, gloss, strength, flexibility, nutritional value, and general performance or coating formulations. Plasticizers have been incorporated into edible coatings as a processing aid to facilitate coating application and to increase merchantability. The most commonly used plasticizers are polyols (sorbitol and glycerol, mono, di or oligo saccharides, lipids).

The plasticizing effect is due to their ability to reduce internal hydrogen bonding while increase intermolecular spacing. This will reduce the brittleness of the film but increases the permeability of water. Several wax formulations have been developed and tasted experimentally. Suitable emulsifiers are used to produce a wax emulsion. Synthetic or natural resins are added to the wax emulsions to give more gloss to the treated produce. Waterwax emulsions are safer to use than solvent waxes, which are highly flammable.

Water -wax emulsion can be used without drying the fruit prior to the treatment, therefore washing and wax treatment can be combined. Carnauba wax, sugarcane wax, thermoplastic terpene resins and shellac are common substances, which are used commercially; however waxes of plant origin are preferred. Development in wax emulsions were not limited to the formulation of new types of wax, but also included the addition of chemical fungicide, which gave waxing treatment an additional function. For example, biphenyl amine added to the wax emulsion not only gave additional shine and reduced weight loss and respiration rate of apples and pears, but also reduced the occurrence of scald.

Addition of post harvest fungicides to the wax, emulsion to control Penicillium rot in stored citrus fruits also proved helpful. Waxes may be applied by dipping, spraying, foaming or brushing. Foaming is a better method of application since it leaves a very thin layer of wax on the fruit. If a thicker layer of wax is applied to the fruit surface, it becomes an undesirable barrier between the external and internal atmosphere and restricts exchange of respiratory gases (CO_2 and O_2). This may result in anaerobic respiration, resulting in fermentation and development of an off flavour. It is therefore necessary to adjust the thickness of the wax coat according to the variety and storage and marketing temperatures.

Effect of Wax Coatings

Wax Coating of fruit improved its external appearance by giving additional shine to the fruit surface. It reduced weight loss and kept the fruit firmer, thus marinating its fresh look, while it reduced rates of respiration and ethylene production, thus delaying senescence. No significant change in constituents like Vitamin C, acidity, and sugar contents (In citrus fruit especially) were recorded due to treatment. However higher concentration of wax resulted in the development of off flavour in citrus fruits particularly during holding at 20°C, making this treatment unsuitable for this fruit. It was, however found necessary to tailor the thickness of the wax coat according to storage temperature to avoid development of off flavour.

FDA regulates these waxes, or coatings, as food additives approved or "generally recognised as safe" for human consumption. However, some consumers have concerns about their use. Vegetarians and others who avoid animal products may worry that fruits and vegetables contain animal-based waxes, such as oleic acid. Some people fear that the wax traps pesticides, making the fruit or vegetable unsafe to eat-even though FDA's pesticide monitoring programme indicates that pesticide residues on fruits and vegetables are consistently within acceptable safe limits.

If consumers want to avoid waxed fruits and vegetables, FDA regulations that took effect in 1994 may help them identify the appropriate products for them. These regulations require produce packers or grocers to provide point-of-sale information about the presence of waxes on fresh fruits and vegetables. This information can appear on labels of individual products, packing cartons (if they are used at the point of sale), or on counter cards or signs. The information will say that the product is: Coated with food-grade animal-based wax to maintain freshness, or Coated with food-grade vegetable-, petroleum-, beeswax-, and/or shellacbased wax or resin to maintain freshness.

If only one of these types of waxes is applied, the label can simply identify the type, such as "vegetable-based." FDA also will allow the statement "No wax or resin coating" on fresh fruits and vegetables that do not contain wax. Besides reading labelling information, consumers can reduce their concerns about waxes by rinsing fruits and vegetables with warm water and, when appropriate, scrubbing with a brush. This will eliminate much of the wax.

5

Drying Fruits and Vegetables

Drying is the oldest known method of preserving food. Historically, food was dried in the sun. Nowadays, we can sun dry or dry in an oven or a dehydrator that is especially designed for home drying. Drying has many advantages. It is safe. By removing most of the moisture from a food, microorganisms can't grow in it. Dried food is compact so that less storage space is needed. Plus, dried foods keep well, travel well and add variety to our diets.

Fruits and vegetables selected for drying should be of the highest quality—fresh and ripe. Drying does not improve the quality. Immature produce will lack flavour and colour. Overmature produce may be fibrous or mushy. If the food is not perfect for eating, it is not suitable for drying.

Dried vegetables can be used as chips, reconstituted for a cooked side dish, used in soups, stews, casseroles and stuffings or made into powders. Dried fruits tend to be chewy and make delicious snacks. Pieces of dried fruits are good in cookies, muffins, cakes and reads. They can also be reconstituted and used in sauces, pies or cobblers or added to gelatin salads, cooked cereals and ice cream.

Different Methods for Drying

Foods can be dried in an electric dehydrator, in the sun, in a solar dryer, or in a regular oven.

Dehydrator Drying

Dehydrator drying produces the best quality dried products, so it's not surprising that it's also the most popular drying method. Dehydrator drying also gives you greater flexibility than other methods because it does not depend on dry, sunny days or take over your oven. A variety of electric dehydrators are available for purchase. A dehydrator should have a heat source, a thermostat, and some method of air circulation. If you buy a dehydrator, follow the directions that come with it. Before you buy a new or used food dehydrator, check to see that it has all these features:

— Instruction manual.

— Thermostatically controlled temperature dial with settings between 130° and 150°F. (If you plan to dry meat jerky in your dehydrator; the dehydrator must be capable of maintaining a temperature of 145°. Contact the extension educator in your county for instructions on how to safely prepare meat jerky.)

— Fan or blower to distribute warm air evenly. Shelves made of stainless steel or food-grade plastic. (Galvanized screening is not food-safe.)

— Easy loading and unloading features.

— Outside cabinets made of hard plastic, aluminum, or steel. The highest quality dehydrator has double-wall construction with insulating material sandwiched between the walls to reduce the amount of heat lost during use.

— Enclosed heating element. Appropriate number of trays for your use. Most food dryers come with 4 to 10 food trays.

— Source of replacement parts.

Sun Drying

Sun drying works best when the temperature is in the 90s, the humidity is low, and air pollution levels are low. A major advantage to sun drying is its low cost. Drying trays, netting to protect against bugs, and food to dry are your only investments. Sun drying makes you dependent on the weather, however. If it is sunny one day and cloudy the next, you will have to finish drying by another method. That's because spoilage can occur while the drying food still has enough moisture for microbial growth. Also, you should bring the food inside on cool nights. Another disadvantage is time. What dries in 6 to 8 hours in an electric dehydrator may take 2 to 4 days in the sun.

Solar Drying

Solar drying is like sun drying, only better. The sun's rays collect in a solar box so that, compared with sun drying, drying temperature is higher and drying time is shorter. The shorter drying time gives microorganisms less chance to cause spoilage. If you do not want to buy or build a solar box, you can use the back window ledge of an automobile where the sun shines through. Crack the windows slightly to allow air flow so temperatures do not get too hot. Cover the trays with netting to keep bugs out. Oven Drying You can use your oven to dry small amounts of food at one time. You'll have little or no investment in equipment and you won't have to depend on the weather. Although oven drying produces a safe, generally tasty product, don't expect top quality. Oven-dried food is more brittle and usually darker and less flavourful than food dried in a dehydrator. Another disadvantage of oven drying is its energy cost.

Oven drying takes two or three times longer than drying in a dehydrator. Before drying in an oven, test the oven temperature with an oven thermometer for about 1 hour. Prop open the oven door as you would when actually drying fruit. The oven should maintain a temperature of 130° to 150°F.If the oven cannot maintain a temperature in this range, you will not have high-quality dried food. If the oven is too hot, your food will begin to cook instead of dry. If it is too cool, your food may not dry fast enough and spoil instead.

If you're new to drying, start with the fruits you like best. Think also about how you will use your dried fruits. Peaches or pears in a tangy stewed fruit? Apples or apricots for lunch box snacks? Berries to toss into muffin or cake batters? Most fruits are easy to dry High-quality fruits make the best dried products. Choose firm, fully ripe fruit that is heavy for its size. Handle fruits gently and process them immediately because fruit ready for drying is very fragile. Use overripe or bruised fruits in other ways (for example, as fruit leathers).

Vegetables for drying should be fresh, tender, and just mature. Avoid immature vegetables because their colour and flavour tend to be weak or poor. Also avoid excessively mature vegetables, which are inclined to be tough, woody, or fibrous. For the best quality and nutrition, dry vegetables as soon as possible after harvest.

Preparation of Dried Fruits

Fruits

Gently wash all fruits in cold water just before drying to

remove dirt, bacteria, and insects. Thoroughly wash fruits that have skins you will not peel off, such as cherries and prunes. Do not soak fruit because extended soaking can cause nutrient loss and waterlog the fruit, which increases drying times. Remove fruit stems and peels. Peels may be left on some fruits, such as apples and peaches, but they may become bitter or discolour during drying. Core or pit the fruit and cut it into uniform halves, quarters, or slices. Trim away diseased or soft spots.

Vegetables

Wash vegetables in cold water just before drying. If vegetables are covered with soil, wash them under clean running water to prevent the dirt from resettling on the food. Do not allow vegetables to soak in water. Most vegetables should be peeled and trimmed then cut, sliced, or shredded into uniform pieces. Although peeling some vegetables such as young zucchini and well-washed carrots is optional, unpeeled vegetables tend to be tougher when dried. Remove fibrous or woody portions and damaged areas.

You can prepare pieces with a food slicer or food processor. Although you can dry and store many foods without pretreatment, pretreatment generally improves quality, particularly for vegetables. Five major reasons for treating foods before drying are to:

- — Preserve colour and flavour
- — Minimise nutrient loss
- — Stop decomposition (enzyme action)
- — Ensure more even drying
- — Extend storage life

Pretreating Fruits and Vegetables

Decomposition from enzyme action during storage is less a problem with fruits than it is with vegetables. Fruits have higher levels of sugar and acid, which counteract enzyme action. Although pretreating fruit is not necessary, you can use an ascorbic acid/citric acid dip, a salt solution dip, syrup blanching, a honey dip, or a sulphiting procedure. Certain fruits, such as apricots, pears, peaches, and some varieties of apples, tend to discolour with drying. Pretreating those fruits can decrease browning during processing and storage and lower losses of flavour and of vitamins A and C.

If you use a pretreatment method that requires soaking fruits in a water solution, you will need to increase drying time because the fruit will absorb some water. Do not allow foods to soak more than 1 hour. Ascorbic Acid/Citric Acid Dips. Ascorbic acid/citric acid dips are often used as a pretreatment for fruits. They prevent fruits such as apples, pears, peaches, and apricots from turning brown when cut and exposed to air. An ascorbic acid dip also increases the vitamin C content of the dried fruit. (Ascorbic acid is another name for vitamin C.)

Use U.S.P. ascorbic acid or food-grade ascorbic acid, which are seasonally available among canning supplies in supermarkets. Vitamin C tablets can also be used. To prepare an ascorbic acid solution, combine ½ teaspoon of ascorbic acid crystals, or three crushed, 500-milligram tablets of vitamin C, with 1 quart water. Stir until the ascorbic acid dissolves. Place the cut fruit in the ascorbic acid solution. Stir the fruit to ensure even coating. Leave the fruit in the ascorbic acid solution for about 5 minutes. Approximately 1 quart of solution will treat 8 cups of fruit.

Pineapple juice or juice from citrus fruits such as oranges, lemons, or grapefruit can also be used as a pretreatment. These juices contain a mixture of citric and ascorbic acids. However, citric acid is a weaker acid than ascorbic acid and is less effective as a pretreatment. You can also use a commercial pretreatment such as the anti-darkening powders often sold with food preservation supplies. Follow the label directions.

Salt Solution Dip. Prepare a solution of 2 to 4 tablespoons of salt per gallon of water. Soak fruit for 2 to 5 minutes, and then drain it well.

Syrup Blanching. Prepare fruit for drying. Prepare a sugar syrup made with 1 part sugar and 2 parts water. If desired, use less sugar. Bring the syrup solution to a boil. Add the fruit, simmer for 5 minutes, then drain the fruit. Place the fruit on drying trays and dry. This fruit product is like a candied fruit.

Honey Dip. A honey treatment for fruit can effectively minimise browning and softening in light-coloured fruit. Prepare a honey-water dip using 1 part honey to 4 parts water. Dip the fruit in the honey solution immediately after slicing, let it soak for about 5 minutes, and drain well. The dried fruit will have a slight honey taste.

Sulphiting. Sulphur dioxide treatments, either sulphiting or sulphuring, are very effective for retarding oxidation and browning in fruit. Fruit flavour and storage life may also improve. Almost all commercially produced light-coloured fruits, such as dried apples, pears, and apricots, are treated with sulphur compounds. However some people have severe allergic responses to sulphur compounds. They should not eat or work with dried fruit pretreated with sulphur or sulphite compounds.

Sulphuring, a complicated and potentially dangerous procedure, is no longer recommended. Sulphiting involves preparing a solution of water and a sulphiting agent and then soaking the cut fruit in the solution. In the United States six sulphur compounds (sulphur dioxide, sodium sulphite, sodium bisulphite, potassium bisulphite, sodium metabisulphite, and potassium metabisulphite) have been listed by the U.S. Food and Drug Administration (FDA) as "Generally Recognised as Safe" (GRAS).

The most popular sulphiting agents for home drying are sodium bisulphite, sodium sulphite, and sodium metabisulphite. They should be either U.S.P. (food grade) or reagent grade (pure). They are available at most wine-making supply centers and some larger supermarkets.

Amount of sulphur to add per quart of water

Sodium bisulphite	½ to 1 teaspoon
Sodium sulphite	1 to 2 teaspoons
Sodium metabisulphite	1 to 3 teaspoons

The sulphiting process has two steps:

- Prepare the sulphiting solution in a large glass container just before use. Place the cut fruit in the solution. Do not leave the fruit in the sulphiting solution too long or the fruit will be mushy. Use about 10 minutes for sliced fruit and 30 minutes for halved fruit. Do not exceed the recommended quantities of sulphites or soak times.
- After sulphiting, remove the fruit and drain it well. Some people recommend a quick rinse in cold water before drying. Place sulphited fruit on drying trays and dry. Drying times for sulphited fruits are longer because the fruit absorbs some water during soaking.

Blanching (heating in boiling water or steam) is the pretreatment method of choice for vegetables. Almost all vegetables should be blanched before drying to destroy the enzymes that make vegetables deteriorate. Blanching keeps vegetables from browning, becoming bitter, or developing off flavours. Blanching also cleans and softens vegetables and makes them easier to rehydrate later. Although you can use either boiling water or steam for blanching, vegetables lose more nutrients during boiling.

Steam Blanching. Use a steamer or make a steamer out of a kettle with a tight-fitting lid. Place a colander, wire basket, or sieve inside the kettle. Make sure the food will be above the water level. Add 2 inches of water to the kettle and heat it to boiling. Place the container with the loosely packed food in the steamer, cover the kettle tightly, and continue boiling.

Water Blanching. Fill a kettle with enough water to cover the food. Bring the water to a rolling boil and gradually stir in the food. Cover the kettle tightly and boil. You can reuse the water when blanching more of the same food, adding more water as necessary. If the water appears dirty, replace it with clean water.

Determining Blanching Times. Blanching times vary with altitude (higher altitudes require longer blanching times), the type and texture of the vegetable, the amount of vegetable, and the thickness of the pieces. Generally, vegetables should feel and taste firm yet tender. They should not be fully cooked, but they should be heated all the way through. Test the food by cutting through a piece. If sufficiently blanched, it will appear cooked (translucent) nearly to the centre. Underblanching may cause deterioration in storage, poor rehydration, or bad colour. Overblanching makes vegetables lose colour, flavour, and nutrients and gives them poor texture after rehydration.

After Blanching. Drain vegetables by pouring them directly on the drying trays. If you plan to reuse the water, place a large pan under the trays. Wipe the bottom of the drying tray with a clean towel to remove excess water. Draining the vegetables on one tray and then transferring them to the drying tray results in unnecessary handling. Immediately transfer the blanched vegetables into the dehydrator so drying can begin while the vegetables are still warm.

Different foods can be dried at the same time, but try to choose foods that will dry in about the same amount of time. (Dry similarly sized pieces together.) Onions, peppers, and other strong foods tend to flavour other foods, so dry them separately. Moisture must be removed from the food as quickly as possible at a temperature that does not seriously impair the flavour, texture, or colour of the food.

If the temperature is too low at the beginning, the food may spoil before it dries. If the temperature is too high, the surface may harden so that the interior dries much more slowly. Start the dryer at 140° to 150°F, with the exceptions noted in the drying guidelines. After 2 to 3 three hours, lower the dryer temperature to 130°F to 140°F. Adequate air flow can reduce drying times. Monitor the drying process. If necessary, rotate the trays to ensure even drying. You may need to stir grated, shredded, or finely cut foods.

Drying Time

Many factors affect drying time, including type of food, size and moisture content of the food pieces, pretreatment method, dryer type, dryer temperature, relative humidity of the air, and amount of air movement in the dryer and in the surroundings. With so many factors at work, it's impossible to give precise drying times. Generally, you can figure on drying times of 6 to 36 hours for fruit and 3 to 16 hours for vegetables, which take less time due to their lower sugar contents.

In the end, you need to decide when food is dry. Vegetables are sufficiently dry when they are brittle or leathery. Leathery vegetables will be pliable and spring back if folded. Brittle vegetables such as corn and peas will shatter when hit with a hammer. Fruits are sufficiently dry when they are pliable and leatherlike and have no pockets of moisture. Herbs are sufficiently dry when brittle. Their leaves will shatter when rubbed together.

When you think the food is sufficiently dry, remove a piece and allow it to cool completely. Then check for dryness. When you are in doubt about the dryness of a food, continue to dry it. Foods dry more quickly toward the end of the drying period, so check them frequently, and avoid leaving them in the dryer after they are done. Leaving them in will reduce their quality.

Packaging

Good packaging and storage techniques are crucial. Packaging protects your dried food from oxygen, moisture (gain or loss), light, microorganisms, and pests. After you have checked foods and found them to be thoroughly dry and cool, pack them immediately for storage.

Conditioning

Some pieces of fruit will be more moist than others after drying so it is a good idea to condition fruits before long-term storage. Conditioning distributes moisture evenly in the fruit. It reduces the chance of spoilage, particularly from mold. To condition, loosely pack cooled, dried fruit in plastic or glass containers to about two-thirds full. Cover the containers tightly. Shake them daily for about 2 to 4 days. The excess moisture in some pieces will be absorbed by the drier pieces. If you notice water forming on the container lid, place the fruit back in the dehydrator. Because vegetables dry to a nearly waterless state, conditioning vegetables is not necessary.

The ideal container for a dried food is:

- Clean and sanitary
- Nontoxic
- Lightweight
- Easily disposable or recyclable
- Moisture resistant
- Airtight
- Protective against light
- Easily opened and closed
- Impermeable to gases and odors
- Durable
- Low-cost

Unfortunately no single food container has all these characteristics. Make your choice based on the type of dried food, your intended storage conditions, and storage time. Three materials-glass, plastics, and metal (never galvanized steel)-are used for packaging most dried foods. Even open-and-close plastic bags are suitable. One good method of storing dried food is to place sealed plastic bags inside a larger glass or metal container with a tight-fitting lid. This twostep packaging has the advantages of being relatively easy, allowing more food to be stored in one container, and protecting against insects and other pests. Although you could store more than one type of dried food inside the larger glass or metal container, do not combine foods with strong odors such as onions, cabbage, or broccoli because other dried foods may absorb their odors.

Labeling

Label each package with the type of food, pretreatment step, and date. Labels may be taped on the outside of a package, tied on with string, or inserted into a clear glass or plastic package. With proper labels you will not have to open individual packages each time you want to use a dried food.

Storage

The length of time you can store dried food depends on

- The type of food
- Factors related to the drying process (pretreatment and final level of moisture in the dried food)
- Packaging of the dried food
- The storage area

An ideal storage area for dried food is cool, dark, and dry. The cooler the storage area, the longer the shelf life. Dark areas are ideal because light fades fruit and vegetables and decreases their vitamin A and C contents. The storage area need not be fancy; a dark, unheated closet or drawer works fine. Metal containers have the advantage of keeping their contents in darkness. Glass or plastic containers can be covered with a cardboard box, a barrel, or black plastic to keep light out. Many people store dried foods in the refrigerator or freezer, which keeps quality high.

During storage at room temperature, the most common type of spoilage is mold growth. Molds can grow in foods that are not completely dry and in foods that absorb water when they are packaged or stored in moist conditions. (Remember: don't consume moldy foods. Some toxic molds can grow at room temperature.) Dried food will probably not absorb enough water to allow bacterial or yeast spoilage.

One typical change that occurs during storage is "Maillard browning," which involves complex chemical reactions between the food's sugars and proteins. Other chemical changes that may take place during storage include loss of vitamin C or other nutrients, general discoloration, changes in food structure leading to an inability of the dried food to fully rehydrate, and toughness in the rehydrated cooked product.

Fruit Leathers

Fruit leathers are a wonderful way to use small quantities of fruit or extra-ripe fruit. Fruit leathers, also known as fruit paper and fruit taffies, are chewy fruit roll-ups made from either cooked or uncooked fruit purée. Fruit leathers allow for individual creativity through combinations of different fruits. Generally you can use any kind of fruit, including apples, bananas, berries, grapes, mangos, papayas, peaches, pears, pineapple, plums, and even tomatoes. Citrus fruits alone are generally not recommended. Fruit leathers are an excellent use for slightly overripe or bruised fruit that would otherwise be discarded.

Uncooked Fruit Leathers

1. Select ripe or overripe fruit or fruit combinations.

2. Wash fruit and cut away blemishes. Remove stones or pits. Remove larger seeds from berries, grapes, and tomatoes if you wish. Peel all tough-skinned fruits; peel others if you wish.
3. Cut fruit into chunks and place them in a food chopper, blender, or food processor.
4. Add 1 tablespoon lemon or other citrus juice per quart of yellow or light-coloured fruit, if desired, for keeping fruit colour.
5. Chop, grind, or blend the fruit into a thick purée. If the fruit has little juice, add several spoonfuls of water or fruit juice to obtain a uniform purée. (If uncooked fruit purée is too juicy, it can be cooked to remove excess liquid.)
6. (Optional) Add sugar, honey, or corn syrup to taste. (Generally no additional sweetener is needed, particularly with ripe fruit, because fruit tastes sweeter after being dried.)
7. (Optional) Add spices (for example, cinnamon, nutmeg, cloves, allspice) to taste. Start with ½ teaspoon dried spice per quart of puréed fruit; spice flavours intensify during drying.
8. Use a drying tray designed for fruit leather or line a portion of a drying tray with lightly oiled heavy plastic wrap. Do not completely cover the tray with plastic wrap or the air will be unable to circulate to other trays. Pour a small amount of purée onto the lining wrap. Make sure the tray has an edge to prevent spillage. Tilt the tray until the purée spreads no more than ¼ inch thick almost to the edge of the plastic wrap. (Two cups of purée will cover a 12- by 17-inch drying tray.)

9. (Optional) Sprinkle the purée with chopped nuts, seeds, or grated coconut.

Cooked Fruit Leathers (Double-Boiler Method)

1. Select, wash, and prepare fruit as described for uncooked fruit leather.
2. Cut the fruit into slices or chunks and place them in the top of a double boiler.
3. Add water to the bottom of the double boiler. Cover the double boiler and steam the fruit for 15 minutes or until fruit is soft. (If a double boiler is not available, you can place a small pan containing the fruit in a larger pan that is partially filled with boiling water.)
4. Follow steps 4-9 for uncooked fruit leather.

Cooked Fruit Leathers (Microwave Method)

1. Select, wash, and prepare fruit as described for uncooked fruit leather.
2. Cut the fruit into slices or chunks and place them in a microwavesafe bowl.
3. Cover the bowl with plastic wrap or a microwave lid.
4. Microwave on high for 4 minutes, stir, and rotate the bowl. Continue stirring and rotating the bowl every 4 minutes until fruit is soft.
5. Follow steps 4-9 for uncooked fruit leather.

Fruit Leather from Canned Fruit

1. Thoroughly drain home-canned or commercially canned fruit or use baby food fruit without tapioca.

2. Follow steps 3 and 5-9 for uncooked fruit leather. Since canned fruits have been heat processed to stop enzymatic action, you don't need to add ascorbic acid.

Drying and Storing

Place prepared trays in the dehydrator. Dry until the leather is sticky, generally 6 to 8 hours at 140°F. Properly dried fruit leather will be translucent and slightly tacky to the touch but will still peel away from the plastic wrap. Lift an edge of the leather, which should stick tightly to the surface, and peel it back slightly. If the leather peels away readily, it is dry. If the leather has cooled, it may need to be warmed slightly for a few minutes to help it peel away.

If the fruit leather cracks or chips, it has dried too long, but it is still edible. Remove the remainder of the plastic wrap. If part of the leather is still sticky, you can dry it more without the plastic wrap. After drying the fruit leather, leave it whole or cut it into pieces. The fruit leather can be rolled and wrapped in plastic wrap or stored flat in sheets with plastic wrap separating each sheet. Place the wrapped pieces in an air-tight container in a cool, dark, dry place. You can also store fruit leather in the refrigerator or freezer.

6

Storage of Fruits and Vegetables

Most fruit and vegetable production in temperate areas is seasonal. In contrast, cultivation and harvest periods are much longer in tropical and subtropical areas. Demand is year round and it is normal practice to use storage in order to ensure continuity of supply. Moreover, storage is a strategy for achieving higher returns. Produce can be held temporarily to overcome gluts thus limiting price falls or to address shortage periods when prices are high. Storage time depends on the intrinsic characteristics and perishability of the product.

Shelf life ranges from short—e.g. raspberries and other berries—to those which naturally adapt to longer storage periods—e.g. onions, potato, garlic, pumpkins, etc. Storage conditions also depend on specific product characteristics. For example, some commodities tolerate temperatures close to 0 °C such as leafy vegetables. Others, such as most tropical fruits (Table 1), cannot tolerate exposure to temperatures below 10 °C.

To optimise storage conditions, not more than one crop should be stored in the same room, unless this is for a short period of time. Sharing the same storage area can result in: differences in temperature and relative humidity conditions; chilling and ethylene sensitivity; odor contamination and other problems affecting shelf life and quality.

Storage Facilities

Generally, storage facilities are linked or integrated to packinghouses or other areas where there is a concentration of product. However, often storage can also be undertaken on-farm, either naturally or in specifically designed facilities. Even under conditions of mechanical refrigeration, location and design have an impact on system operations and efficiency. First, climate is an important factor for the location of the storage facility. For example, altitude reduces temperature by 10 °C for every 1000 meters of elevation. It also increases overall efficiency of the refrigeration equipment by facilitating heat exchange with ambient temperature, thereby reducing energy costs. Shading particularly of loading and unloading areas reduces thermal differences between field and storage temperatures.

Table 1. Recommended temperature, relative humidity and storage life for fruits and vegetables

	Crop temperature (°c)	Relative Humidity (%)	Storage Life (days)
Amaranth	0-2	95-100	10-14
Apple	-1-4	90-95	30-180
Apricot	-0.5-0	90-95	7-21
Artichoke	0	95-100	14-21
Asian pear	1	90-95	150-180
Asparagus	0-2	95-100	14-21
Atemoya	13	85-90	28-42
Avocado	3-13	85-90	14-56
Babaco	7	85-90	7-21
Banana—Plantain	13-15	90-95	7-28

Contd....

Contd....

Barbados cherry	0	85-90	49-56
Basil	7-10	85-95	7
Bean (dry)	4-10	40-50	180-300
Beet (bunched)	0	98-100	10-14
Beet (topped)	0	98-100	120-180
Belgian endive	0-3	95-98	14-28
Blackberry	-0.5-0	90-95	2-3
Black sapote	13-15	85-90	14-21
Blueberries	-0.5-0	90-95	14
Bok Choy	0	95-100	21
Breadfruit	13-15	85-90	14-42
Broadbeans	0-2	90-98	7-14
Broccoli	0	95-100	14-21
Brussels sprouts	0	95-100	21-35
Cabbage	0	98-100	150-180
Cactus leaves	2-4	90-95	14-21
Caimito	3	90	21
Calamondin	9-10	90	14
Cantaloupe (half slip)	2-5	95	15
Cantalupo (full slip)	0-2	95	5-14
Carambola	9-10	85-90	21-28
Carrot (bunched)	0	95-100	14
Carrot (topped)	0	98-100	210-270
Cassava	0-5	85-96	30-60
Cashew apple	0-2	85-90	35
Cauliflower	0	95-98	21-28
Celery	0	98-100	30-90
Celeriac	0	97-99	180-240
Chayote	7	85-90	28-42
Cherimoya	13	90-95	14-28
Cherries	-1-0.5	90-95	14-21
Chicory	0	95-100	14-21
Chinese cabbage	0	95-100	60-90
Chives	0	95-100	14-21
Coconut	0-1.5	80-85	30-60

Contd....

Contd....

Cranberries	2-4	90-95	60-120
Cucumber	10-13	95	10-14
Currants	-0,5-0	90-95	7-28
Custard apple	5-7	85-90	28-42
Daikon	0-1	95-100	120
Dates	-18-0	75	180-360
Durian	4-6	85-90	42-56
Eggplant	8-12	90-95	7
Escarole	0	95-100	14-21
Fennel	0-2	90-95	14-21
Feijoa	5-10	90	14-21
Fig	-0.5-0	85-90	7-10
Garlic	0	65-70	180-210
Ginger	13	65	180
Grape	-0.5-0	90-95	14-56
Grapefruit	10-15	85-90	42-56
Green onions	0	95-100	21-28
Guanabana	13	85-90	7-14
Guava	5-10	90	14-21
Horseradish	-1-0	98-100	300-360
Husk tomato	13-15	85-90	21
Jaboticaba	13-15	90-95	2-3
Jackfruit	13	85-90	14-42
Jerusalem artichoke	-0.5-0	90-95	120-150
Jicama	13-18	65-70	30-60
Kale	0	95-100	10-14
Kiwano	10-15	90	180
Kiwifruit	-0.5-0	90-95	90-150
Kohlrabi	0	98-100	60-90
Kumquat	4	90-95	14-28
Leek	0	95-100	60-90
Lemon	10-13	85-90	30-180
Lettuce	0-2	98-100	14-21
Lima bean	3-5	95	5-7
Lime	9-10	85-90	42-56

Contd....

Contd....

Longan	1-2	90-95	21-35
Loquat	0	90	21
Lychee	1-2	90-95	21-35
Malanga	7	70-80	90
Mamey	13-18	85-95	14-42
Mandarin	4-7	90-95	14-28
Mango	13	90-95	14-21
Mangosteen	13	85-90	14-28
Melon (Others)	7-10	90-95	12-21
Mushrooms	0-1.5	95	5-7
Nectarine	-0.5-0	90-95	14-28
Okra	7-10	90-95	7-10
Onions (dry)	0	65-70	30-240
Olives, fresh	5-10	85-90	28-42
Orange	0-9	85-90	56-84
Papaya	7-13	85-90	7-21
Parsley	0	95-100	30-60
Parsnip	0	95-100	120-180
Passionfruit	7-10	85-90	21-35
Peach	-0.5-0	90-95	14-28
Pear	-1.5-0.5	90-95	60-210
Peas	0	95-98	7-14
Cucumber	5-10	95	28
Pepper (bell)	7-13	90-95	14-21
Persimmon	-1	90	90-120
Pineapple	7-13	85-90	14-28
Pitaya	6-8	85-95	14-21
Plum	-0.5-0	90-95	14-35
Pomegranate	5	90-95	60-90
Potato (early)	7-16	90-95	10-14
Potato (late)	4.5-13	90-95	150-300
Prickly pear	2-4	90-95	21
Pumpkins	10-15	50-70	60-160
Quince	-0.5-0	90	60-90
Radichio	0-1	95-100	14-21

Contd....

Contd....

Radish	0	95-100	21-28
Rambutan	10-12	90-95	7-21
Raspberries	-0.5-0	90-95	2-3
Rhubarb	0	95-100	14-28
Rutabaga	0	98-100	120-180
Salsify	0	95-100	60-120
Sapodillla	15-20	85-90	14-21
Scorzonera	0	95-98	180
Snapbeans	4-7	95	7-10
Snowpeas	0-1	90-95	7-14
Spinach	0	95-100	10-14
Sprouts	0	95-100	7
Strawberry	0-0.5	90-95	5-7
Sweet corn	0-1.5	95-98	5-8
Sweet potato	13-15	85-90	120-210
Swiss chard	0	95-100	10-14
Summer squash	5-10	95	7-14
Tamarind	7	90-95	21-28
Taro	7-10	85-90	120-150
Tart cherries	0	90-95	3-7
Tomato (MG)	12.5-15	90-95	14-21
Tomato (red)	8-10	90-95	8-10
Tree tomato	3-4	85-90	21-28
Turnip	0	90-95	120
Watercress	0	95-100	14-21
Watermelon	10-15	90	14-21
White sapote	19-21	85-90	14-21
Yam	16	70-80	60-210
Yellow sapote	13-15	85-90	21

Building design is an important factor to be taken into consideration. For example, a square shaped floor perimeter is thermally more efficient than a rectangular one. The roof is the most important part of the structure. This is because it has to protect produce from rain and radiant heat. Its slope should allow easy fall off of rainwater; its dimensions should exceed the perimeter of the building to protect walls from the sun and provide a dry area around the building in rainy weather. Floors should be of concrete, isolated from soil humidity, and elevated to avoid penetration of water.

Doors need to be wide enough for mechanised handling. Storage facilities should be thoroughly cleaned before filling. This includes brushing and washing of walls and floors to eliminate dirt and organic debris that could harbour insects and diseases. Before product is placed in the storage room, inspection and presorting should be undertaken. This is in order to remove all potential sources of contamination for the remaining load. Product should be stacked in such a way that there is free circulation of air. During storage, it should also be possible to carry out quality control inspections. If the storage facility becomes full during a long harvest period, it needs to be organised around the principles of the system "first in first out".

Different Ways of Storing

As a rule, there are many ways of storing a product. The length of storage time can be longer in specifically designed structures. With refrigeration and controlled atmospheres, storage periods can be even longer. The technology utilised depends on whether the benefits (higher prices) outweigh the costs.

Natural Storage

This is the most rudimentary system and is still in use for many crops. For example, roots (carrots, sweet potato, and cassava) and tubers (potato). Crops should be left in the soil until preparation for the market. This is similar to how citrus and some other fruits are left on the tree. Although storing products under natural conditions is widely practiced, it leaves them exposed to pests and diseases as well as to adverse weather conditions. This can have a detrimental effect on quality.

Another method widely used is field storage in heaps. This method ensures that produce is free from soil humidity and is protected from the weather with a tarpaulin, straw, or plastic materials. It is a low cost alternative for bulky crops that require large buildings. For example, potato, onions, pumpkins, sweet potato etc. Field storage in bins is a more recent variation where a pair of them (one on top of the other, the one above protected from the weather) is left in the field. It has the additional advantage of making it possible to undertake mechanical handling later.

Natural Ventilation

Amongst the wide range of storage systems, this is the most simple. It takes advantage of the natural airflow around the product to remove heat and humidity generated by respiration. Buildings providing some form of protection from the external environment and with gaps for ventilation can be used. Produce can be placed in bulk, bags, boxes, bins, pallets etc. Although simple, some key concepts need to be taken into account for the efficient operation of this system.

1. Differences in internal temperature and relative humidity conditions compared to conditions externally, need to be minimal. What this means is that this system can only be used with crops that store well under natural conditions such as potato, onions, sweet potato, garlic, pumpkins, etc.

2. For adequate ventilation, openings need to be wide. This means they need to be fitted with screens to keep animals, rodents, and pests out.

3. As with any other type of fluid, air follows the path of least resistance. This means that if product is stored in a compact mass, air will circulate to remove heat and gases which have accumulated as a result of respiration. Efficient ventilation requires adequate space. However, this reduces storage capacity.

4. Hot and humid air rises within the storage facility. If no ventilation gaps exist, this leads to the build up of hot and humid areas which in turn affects the quality of stored goods. This presents the ideal conditions for the development of disease.

Within certain limits, it is possible to take advantage of natural changes in temperature and relative humidity. This can be achieved by selectively opening and closing the storage ventilation. At noon, ambient temperature and relative humidity are higher and lower, respectively. However, at night the opposite happens. To reduce temperature of stored products, buildings should be left open when external air temperatures are lower. Internal relative humidity can also be managed in a similar way.

External conditions constantly change, even during the same day. However, in comparison to air, stored mass is slower to gain and release heat. In order to handle this efficiently, internal and external electronic sensors for temperature and relative humidity are required. In addition to this, although crops suitable for this type of storage have low respiratory rates, some ventilation may be required. This is in addition to the automated opening and closing schedules.

Forced Air Ventilation

Heat and gas exchange can be improved provided air is forced to pass through the stored product. This system allows for more efficient utilisation of space for bulk storage. Air conducts run under a perforated floor and air is forced through the product. Again, as air follows the least resistance path, loading patterns as well as fan capacity and conduct dimensions should be carefully calculated. This is to ensure that there is uniform distribution of air throughout the product. Removable perforated ducts can be used for storage space when there are no products in storage.

Fan selection is the most critical factor and specialised personnel should design the system based on volume and number of air changes per unit of time required. The latter is a function of respiratory rates of products to be stored. Static pressure or resistance to the airflow by conducts and stored mass should be considered. Ideally, sensors reacting to the internal/external ambient relationship should control the system. If closed, internal air circulation only occurs. On the other hand, if opened internal atmosphere is replaced by ventilation. A partial opening produces a mix of internal and external air to reach the desired combination of temperature and relative humidity.

Refrigeration

Controlling temperature is one of the main tools for extending postharvest life: low temperatures slow product metabolism and the activity of microorganisms responsible for quality deterioration. As a result, reserves are maintained with a lower respiration rate, ripening is retarded and vapor pressure between product and ambient is minimised, reducing water loss. These factors contribute towards maintaining freshness by reducing the rate at which quality deteriorates and the nutritional value of the product is preserved. A refrigerated room is a relatively airtight and thermally insulated building.

The refrigeration equipment should have an external escape outlet to release externally the heat generated by the product. Refrigeration capacity of the equipment should be adequate to extract the heat generated by crops with a high respiration rate. It is also important to precisely control temperature and relative humidity conditions inside the refrigerated storage environment. Refrigerated space depends on the maximum storage volume.

Other factors to be considered include walkways and aisles to handle the product mechanically and the additional space to ensure uniform distribution of cold air. It is not uncommon to find that produce occupies only 75-80% of total surface area. Chamber height depends on product and stacking pattern: three meters for hand stacking but more than six may be required if forklifts are utilised.

Refrigerated rooms can be made with concrete, metal, wood, or other materials. All external surfaces should be thermally insulated, including the floor and ceilings. Type and thickness of insulation material depends on building characteristics, produce to be stored and the difference in temperature required between external and internal conditions. Polyurethane, expanded polystyrene, cork and other such materials can be used as insulation materials. A vapor barrier should be placed on the warm side of the insulation material.

Mechanical refrigeration has two main components: the evaporator, inside the storage area and the condenser which is outside connected by tubing filled with refrigerant. Normally, both elements are finned coils made of high thermal conductivity materials and integrated to a fan. This facilitates heat exchange. An evaporator is placed in the upper part of one of the walls forcing cold air to flow parallel to the ceiling. Returning air is forced past the evaporator transferring to the coil the heat extracted from the product.

A refrigerant absorbs this heat as it changes to gas, cooling the air, which is forced again into the room as cold air. The refrigerant is transported as gas to the condenser (outside of building) where under the pressure provided by a compressor, it is transformed again into the liquid form. The internal heat is then released outside. With this repeated cycle, the system behaves like a pump—heat is extracted from the stored product and then released outside. Another key aspect of the mechanical refrigeration system is the expansion valve, which regulates the evaporation and flow of refrigerant.

Ammonia and Freon gas are the most widely used refrigerants. However, they are now being replaced by more environmentally friendly products. In addition to design and consideration of building materials, to gain maximum benefit from refrigeration the following conditions need to be met: refrigeration capacity needs to be adequate—this is in order to extract respiration heat from the product as well as conductive heat (through floors, walls, and ceiling); convective heat gains (door openings), and the heat produced by equipment (forklifts, lights, pumps, etc.).

Every crop has an optimal combination of temperature and relative humidity for storage. In many cases, there are differences even within varieties (Table 1). As previously mentioned, it is recommended not to store more than one crop in the same room, unless this is for a very short period (less than a week) or during transportation. Very incompatible crops should not be in the same room for more than 1 or 2 days.

Precooling

Refrigeration equipment is designed to keep product chilled. However, they are not capable of reducing field heat rapidly. Field temperature is close to the ambient one and is much higher if produce is not protected from the sun. When produce is exposed to colder ambient conditions, it loses field temperature only slowly. It may take up to 24 or 48 hours in order to reach the new ambient temperature. The rate at which temperature falls depends on a number of factors. These include: differences in temperature, individual volume of product, total mass required for precooling and capacity of the refrigeration equipment.

Metabolic activity (respiration, ethylene production, biochemical, and enzymatic reactions) also decreases with temperature—when storage temperature is reached rapidly, this results in reduced losses in energy, stored reserves, and quality. Precooling is the rapid reduction of field temperature prior to processing, storage, or refrigerated transport. Generally it is a separate operation requiring special facilities, but complementary to cold storage. As deterioration is proportional to the time produce is exposed to high temperatures, precooling is beneficial even when produce returns later to ambient conditions. It is critical in maintaining quality in fruits and vegetables and forms part of the "cold chain" to maximise postharvest life.

Product temperature loss is not linear. This is because it is rapid at the beginning but slows down as it approaches the medium refrigerating temperature.

Operation costs increase for each degree reduced. In commercial operations, produce is precooled to reach 7/8th of the difference between field and the final temperature required. The remaining 1/8th is lost during refrigerated storage or transport. For example, a product precooled with a field temperature of 30°C followed by exposure to a refrigerating medium of 10°C, should be terminated when 7/8th of the temperature difference is removed (final temperature = 12.5°C)

Tfinal = Tinitial product - [7 x (Tinitial product - Trefrigerant)/8]

Tfinal = 30 - [7 x (30 - 10)/8] = 12.5°C

The rate of cooling depends on individual volume and the exposed surface of product. The difference in temperature between product and the refrigerating medium also needs to be taken into account. For example, due to large exposed surfaces, leafy vegetables cool almost 5 times faster than large fruit such as melons or watermelons. Other factors which have an influence include the type of cooling medium and the amount of circulation surrounding the product. Water has more capacity to absorb heat than air and rapid circulation increases their cooling capacity. Each system listed below has its advantages and disadvantages.

a. Cold air:	Room cooling
	Forced air cooling
b. Cold water:	Hydrocooling
c. Contact with ice:	Crushed ice
	Liquid ice
	Dry ice

d. Evaporation of surface water: Evaporative

Vacuum cooling

Room cooling: This is probably the most widely used system and is based on the product's exposure to cold air inside a refrigerated room. It is simple to operate as the product is cooled and stored in the same room. However, the removal of heat slowly makes this system unsuitable for highly perishable commodities. This is because the product needs at least 24 hours to reach the required storage temperature. Almost all crops are suitable for this type of cooling but it is mainly used in potato, onions, garlic, citrus, etc. (Table 2).

Table 2. Crops usually room cooled.

Artichoke	Coconut	Melons	Salsify
Asian pear	Custard apple	Onion	Sapote
Atemoya	Garlic	Orange	Scorzonera
Banana	Ginger	Parsnip	Summer squash
Beans (dry)	Grapefruit	Cucumber	Sweet potato
Beet	Horseradish	Pineapple	Tangerine
Breadfruit	Husk tomato	Plantain	Tomato
Cabbage	J. artichoke	Potato	Tree tomato
Cactus leaves	Jicama	Prickly pear	Turnip
Carambola	Kiwano	Pumpkin	Watermelon
Cassava	Kohlrabi	Quince	Yam
Celeriac	Kumquat	Radish	
Chayote	Lime	Rhubarb	
Cherimoya	Lemon	Rutabaga	

Forced air cooling: This system includes cold air being forced to pass through produce by means of a pressure gradient across packages. Cooling is 4 to 10 times more rapid than room cooling and its rate depends on airflow and the individual volume of produce. Amongst the wide range of systems available, this is probably the most versatile. This is because it can be applied to all crops (Table 3), particularly berries, ripe tomatoes, bell peppers and many other fruits.

It is slow compared to hydrocooling but is a good alternative for crops requiring rapid heat removal which cannot tolerate wetting or chlorine of cooling water. However, inadequate airflow may produce dehydration. Package ventilation openings should be large enough to allow adequate air flow, particularly if products are stacked or palletised. Adequate airflow is necessary. This is because fruits in the centre of packages tend to lose heat at a slower rate, compared to those on the exterior.

Table 3. Crops usually precooled by forced air.

Anona	Coconut	Mango	Prickly pear
Atemoya	Cucumber	Mangosteen	Pumpkin
Avocado	Eggplant	Melons	Quince
Banana	Feijoa	Mushrooms	Rhubarb
Barbados cherry	Fig	Okra	Sapote
Berries	Ginger	Orange	Snapbeans
Breadfruit	Grape	Papaya	Snowpeas
Brussels sprouts	Grapefruit	Passionfruit	Strawberry
Cactus leaves	Guava	Pepino	Summer squash
Caimito	Husk tomato	Pepper (Bell)	Tangerine
Carambola	Kiwifruit	Persimmon	Tomato
Cassava	Kumquat	Pineapple	Tree tomato
Chayote	Lima bean	Plantain	Yam
Cherimoya	Lychee	Pomegranate	

Hydrocooling: The refrigerating medium is cold water. Because of its higher capacity to absorb heat, it is faster than forced air cooling. Hydrocooling can be achieved by immersion or through means of a chilled water shower. In this final system, produce must be arranged in thin layers for uniform cooling. Not all crops can be hydrocooled. This is because they need to be able to tolerate wetting, chlorine, and water infiltration. Tomato, asparagus and many other vegetables are hydrocooled commercially (Table 4). Chlorination of water (150-200 ppm) is important to prevent accumulation of pathogens.

Ice cooling: This is probably one of the oldest ways to reduce field temperature. The most common method of ice cooling is at the individual pack level—crushed ice is added to the top of the product before the package is closed. Ice layers may also be interspersed with produce. As it melts, cold water cools the lower layers of product. Liquid icing is another system where a mix of water and crushed ice (40% water + 60% ice + 0,1% salt) is injected into open containers so that a big ice block is formed.

Table 4. Crops normally hydrocooled.

Artichoke	Cassava	Kiwifruit	Radish
Asparagus	Celeriac	Kohlrabi	Rhubarb
Beet	Celery	Leek	Salsify
Belgian endive	Chinese cabbage		Lima bean
Snapbeans			
Broccoli	Cucumber	Orange	Snowpeas
Brussels sprouts	Eggplant	Parsley	Spinach
Caimito	Escarole	Parsnip	Summer squash
Cantaloupe	Green onions	Peas	Sweet corn
Cauliflower	Horseradish	Pomegranate	Swiss chard
Carrot	J. artichoke	Potato (early)	Watercress

The main disadvantage of ice cooling is that it is limited to ice tolerant crops (Table 5). It also increases costs because of the heavier weight for transportation and the need for oversized packages. In addition to this, as water melts, storage areas, containers, and shelves become wet.

Table 5. Crops that can be ice cooled.

Belgian endive	Chinese cabbage	Kohlrabi	Spinach
Broccoli	Carrot	Leek	Sweet corn
Brussels sprouts	Escarole	Parsley	Swiss chard
Cantaloupe	Green onions	Pea/snowpeas	Watercress

Evaporative: This is one of the most simple cooling systems. It involves forcing dry air through wet product. Heat is absorbed from product as water evaporates. This method has a low energy cost but cooling efficiency is limited by the capacity of air to absorb humidity. As a result, it is only useful in areas of very low relative humidity.

Table 6. Crops that can be vacuum cooled.

Belgian endive	Celery	Mushrooms	Sweet corn
Brussels sprouts	Escarole	Radiccio	Swiss chard
Carrot	Leek	Snapbeans	Watercress
Cauliflower	Lettuce	Snowpeas	
Chinese cabbage	Lima bean	Spinach	

Vacuum cooling: Is one of the more rapid cooling systems. However, this is accomplished at very low pressures. At a normal pressure of 760 mmHg, water evaporates at 100 °C, but it does at 1 °C if pressure is reduced to 5 mmHg. Product is placed in sealed containers where vacuum is performed. Vacuum cooling produces about 1% product weight loss for each 5 °C of temperature reduction. Modern vacuum coolers add water as a fine spray in the form of pressure drops. Similar to the evaporation method, this system is in general appropriate for leafy vegetables. This is because of their high surface-to-mass ratio (Table 6).

Chilling injury

Refrigeration is the most widely used method for extending the postharvest life of fruits and vegetables. However, low temperatures may produce injuries to plant tissues. Freezing (prolonged exposure to temperatures lower than 0°C); forms ice crystals inside tissues. This causes damage. Symptoms are readily apparent when thawing occurs—there is loss of turgidity and a general breakdown of plant tissues. One of the main causes for this injury is unattended or malfunctioning refrigeration equipment.

Chilling injury on crops that do not tolerate long exposure to temperatures in the range of 0 - 15°C are less noticeable. Most chilling sensitive crops are of tropical or subtropical origin. For example, tomatoes, peppers, eggplants, pumpkins, summer squash, sweet potato, banana etc. Some temperate crops may also be sensitive. For example, asparagus, potato, some apple varieties, peaches etc. Critical temperatures for these crops range from 0-5°C, while those of tropical origin are from 7-15°C. Symptoms of chilling injury depend on the type of crop and become noticeable when product is returned to ambient temperature.

In banana, for example, a blackening of the skin and softening takes place while in tomato, pepper, eggplants and other fruits, sunken areas are apparent. This is usually associated with decay organisms and followed by rapid and uneven ripening. In many cases internal darkening or other discolourations are present. Severity of chilling injury depends on crop, temperature and length of exposure. As a general rule, immature fruits are more susceptible to damage than mature ones. From a physiological point of view, chilling injury is the result of a cumulative breakdown of cellular metabolism. This is reversible during the first phase.

A small rise in temperature restores the product to its former condition provided injuries are of a temporary nature. Different studies show that periodic (from 6-7 to 15 days) and short interruptions (5 to 48 hours) of cold storage through increases in temperature (from 12 to 25°C) contribute towards extending post harvest life. Chilling injury is cumulative. In many cases it is the result of field, storage, and/or low transport temperatures.

Gaseous contamination

Under relatively airtight storage conditions, metabolic gases accumulate and ethylene and other volatiles are some of the most frequent contaminants.

Ethylene is a fitohormone, which regulates many growing, development and senescence processes in plant tissues. It is produced in large quantities by climacteric fruits during ripening. It is also induced by certain types of stress such as physical injuries and is also part of the healing process. Ethylene is released as a gas and accumulates to physiologically active levels when not eliminated by ventilation or chemical means. When ethylene releasing and sensitive (Table 7) crops are placed in the same room, undesirable reactions take place. For example, these include an increase in respiratory rate, ripening and senescence, loss of green colour, yellowing, necrotic areas on plant tissues, formation of abscission layers, sprouting in potatoes, development of bitter flavour in roots, asparagus toughening, etc.

Table 7. Ethylene and odor producers and sensitives.

Ethylene producer	Ethylene sensitive	Odor producer	Odor sitive
Anona	X	X	
Apple	X	X	XX
Apricot	X	X	
Asian pear	X	X	
Asparagus		X	
Atemoya	X	X	
Avocado	X	X	XX
Banana	X	X	
Basil		X	
Belgian endive		X	
Broccoli		X	
Brussels sprouts		X	
Cabbage		X	X
Cactus leaves		X	
Carrot		X	XX
Cauliflower		X	
Celery		X	X
Cherimoya	X	X	
Cherry			X
Chinese cabbage		X	
Chives		X	
Cucumber		X	
Dates			X
Eggplant		X	X
Escarole		X	
Feijoa	X		
Fig	X		X
Grape			XX
Green onions		X	X
Guava	X	X	
Husk tomato		X	
Jackfruit	X	X	

Contd....

Contd....

Kale		X	
Kiwano		X	
Kiwifruit	X	X	
Leek		X	X
Lemon			X
Lettuce		X	
Lima bean		X	
Lime			X
Lychee	X	X	
Mandarin		X	
Mango	X	X	
Melons	X	X	
Mushrooms	X	X	X
Nectarines	X	X	
Okra		X	
Olives, fresh		X	
Onion			XX
Oranges		X	X
Papaya	X		
Parsley		X	
Parsnip		X	
Passionfruit	X	X	
Pea		X	
Peach	X	X	
Pear	X	X	XX
Pepino		X	
Pepper (Bell)		X	X
Persimmon	X	X	
Prickly pear		X	
Pineapple			X
Plum	X	X	
Potato		X	XX
Quince	X	X	
Rambutan	X	X	
Sapodilla	X	X	

Contd....

Contd....

Sapote	X	X	
Snapbeans		X	X
Snowpeas		X	
Spinach		X	
Summer squash		X	
Sweet corn			X
Sweet potato		X	
Swiss chard		X	
Tomato	X	X	
Watercress		X	
Watermelon		X	
Yam		X	

Indirect effects include an increase in sensitivity to chilling, or susceptibility to pathogens and the stimulation of some decay organisms. The level of ethylene in storage areas should be less than 1 ppm to avoid problems. Aroma, odors, and other volatiles form an integral part of the metabolism of the plant. As with ethylene, there is contamination when producing species and sensitive crops share the same storage area (Table 7).

Relative humidity

Fruits and vegetables are largely composed of water. An important factor in maintaining post harvest quality is to ensure that there is adequate relative humidity inside the storage area. Water loss or dehydration means a loss in fresh weight. This in turn affects the appearance, texture, and in some cases the flavour. Water loss also affects crispiness and firmness. Consumers tend to demand and associate these qualities with freshness, perceiving them as just harvested.

The percentage of relative humidity is the most widely used parameter to express the amount of water in the air. It is defined as the relationship between the pressure of water in the air and the temperature at saturation point. As with other gases, water vapour moves from higher to lower pressure areas. In plant tissues, water is mainly present as cellular liquids, but in equilibrium with the intercellular spaces where it exists as a vapor saturated atmosphere (100% relative humidity). Exposure to identical air conditions of relative humidity and temperature will prevent water loss from tissues.

Capacity of air to hold water increases with temperature. The reverse is also true. This means that refrigeration increases the relative humidity of air. However, in some cases humidifiers are needed to increase the moisture content so as to reach the ideal conditions for storage. Onion, garlic, pumpkin etc provide some exceptions. They are best stored at relative humidity in the range of 60-70%. Most fruits and vegetables are required to be kept at a relative humidity of 90-95%, while some others at values close to saturation (Table 1).

Short term storage

Refrigeration in cold stores is not always used to maximise the postharvest life. On the contrary, it is probably used more often during the short time required for the sequence of activites in the cold chain ending at the consumption point. Refrigerated transport is probably the best example of this. However, there are many other opportunities for the use of temporary cold storage e.g. during preparation and sale of product for the market. For example, holding product until processing, packaging, or transport is carried out. Other examples include the use of refrigerated facilities at wholesale or retail.

Cold storage is also used in the home to prolong the shelf-life of products. It is hard to define what constitutes "short term and long-term storage". This is because 7 days is a long time for raspberries while for potato, onion, garlic and other products that require longer periods of storage this is considered to be relatively short.

It is preferable not to store different crops together. However, this is common practice and is unavoidable in many cases, particularly at distribution or retail. This does not pose a problem provided products are not exposed to suboptimal conditions for too long and build up of ethylene is avoided. A strategy widely practiced is to set cold chambers at an average of around 5°C and 90-95% level of relative humidity. If possible, mixed loads should have different regimes depending on the specific combination of fruits and vegetables in store. This is assuming that ambient ethylene concentration does not exceed 1 ppm.

Transport is an example of temporary refrigerated storage. Mixed loads cause incompatibility problems highlighted previously. Because packaging dimensions are different, they are usually not fully stackable and the ventilation openings of packages of different dimensions do not match to each other. This prevents ventilation and creates microambient conditions which are undesirable.

Combination of Storage

Facilities for long-term storage of potato, onion, sweet potato etc, often include using a combination of forced air systems as well as heating and/or refrigeration equipment. Because these are crops that initially require a curing period, hot and humid air is introduced at the beginning. Later, the temperature is reduced with either through forced air cooling or natural ventilation.

Adequate temperatures are obtained by mixing external and internal atmospheres and if required, the air is heated or refrigerated. In this way, the same building is used for both curing and storage—an important consideration in mechanised harvesting systems.

Controlled Atmospheres

With atmosphere modification, the low metabolic rate achieved with refrigeration is extended even further. As a result, the storage period is prolonged without further losses in quality. Composition of normal atmosphere at sea level is around 78,1% nitrogen, 21% oxygen y 0,03% carbon dioxide. A "controlled" or "modified" atmosphere is obtained when its composition varies from the norm.

In controlled atmosphere, gas composition is exactly maintained. It is often used for extremely long periods of storage in purpose built facilities. Modified atmospheres, on the other hand, are obtained when produce is packed in semi permeable films and are used for short periods. The atmospheric composition inside the package changes until it is in equilibrium with the ambient one. Equilibrium atmosphere depends on product, film characteristics, and storage temperature.

The modification of storage atmosphere delays the biochemical and physiological changes associated with senescence. This mainly involves the respiratory rate, ethylene production, softening and compositional changes. Other effects include the reduction in sensitivity to ethylene, and in some cases chilling and the severity of pathogen attack. The atmospheric composition can also be used to control insects. The risk of using abnormal atmospheres is that they may cause fermentation, tissue asphyxia, and the development of off-odors or off-flavours. From the construction point of view, controlled atmosphere facilities are similar to refrigeration facilities.

However, they should be airtight to allow creation of an atmosphere different from normal. The Oxygen consumption and its replacement by carbon dioxide by respiration, create the atmosphere. When the appropriate combination has been reached, a limited intake of oxygen is required to satisfy the reduced rate of respiration. Accumulation of carbon dioxide is removed by means of different methods. Because internal atmosphere behaves differently, a pressure compensating system is required to attain equilibrium with the external or ambient atmosphere.

As controlled atmosphere rooms are kept locked until the end of the storage period, inspection windows are required to control refrigeration equipment. Product should also be placed at the top of one of the walls. Atmospheric composition is crop specific. However, as a general rule the most common combinations are 2-5% oxygen and 3-10% carbon dioxide. Many crops benefit from atmosphere modification.

However, usage is limited. It is difficult to define products ideal for storing under controlled atmosphere. However, one of the most important factors is that investment and operating costs should be recovered. Other factors include: First, products should be seasonal and have a stable demand during a long marketing period. Second, product should have some unique qualities and not be easily substituted by similar products. In other words, it is beneficial to use controlled atmosphere technology when there are no competitor products on the market. This may go some way towards explaining why its usage is limited to specific crops, particularly apples and pears.

Home Storage System

Outdoor Storage

Produce that requires cool-to-cold moist surroundings can be stored outdoors. All outdoor storage has the disadvantage of being unaccessible sometimes and subject to damage by rodents and other vermin. A welldrained location is essential to prevent excessive accumulation of water. Usually the produce must be insulated for protection from frost and fluctuating temperatures. Insulating materials commonly used are straw, hay, dry leaves, corn stalks, or wood shavings, and some soil. Be sure that the insulating materials used are not contaminated with pesticides.

In-garden storage

It is possible to leave some root crops, such as carrots, turnips, and parsnips in the garden where they grew, for part or all of the winter. (fig. 1.) After the ground begins to freeze in the late fall, cover the root crops with a foot or more of mulch-straw, hay, or dry leaves. Do not place mulch on warm soil because doing so will cause vegetables to decay rapidly. Wait until the ground is cold.

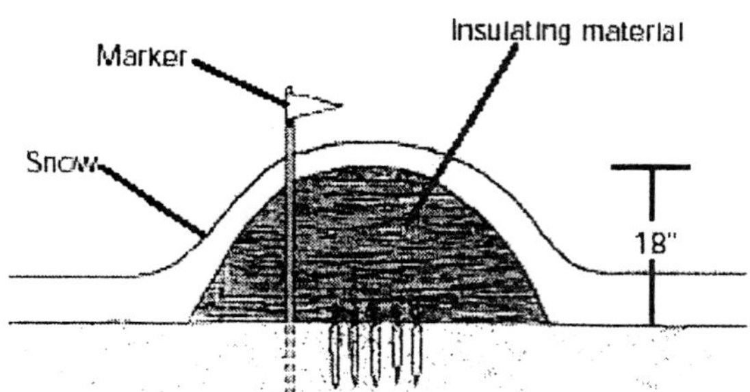

Figure 1. In-garden storage.

Produce can be difficult to dig out of the frozen ground, but it will not be adversely affected until the temperature around the roots drops to 25° or less. Carrots are damaged at about 25°, but parsnips can stand somewhat lower temperatures. If rodents are a problem, it may be wise to store produce in a buried container or an indoor storage area. One gopher can consume a whole row of carrots left in the ground. Parsnips, horseradish, and turnips actually improve in flavour by light freezing. At temperatures between 28° and 34°, the starch changes to sugar. Other crops, such as beets, cabbage, Chinese cabbage, cauliflower, celery, endive, cos or romaine lettuce, kale, leeks, and onions can withstand the early light frosts and can be stored for several weeks under a heavy mulch.

Mounds, pits

Mounds or pits are a very economical way to store cabbage and root crops, such as carrots, beets, celeriac, kohlrabi, rutabagas, turnips, and winter radishes. (fig. 2.) Select a well-drained location, and cover the ground with an insulating mulch.

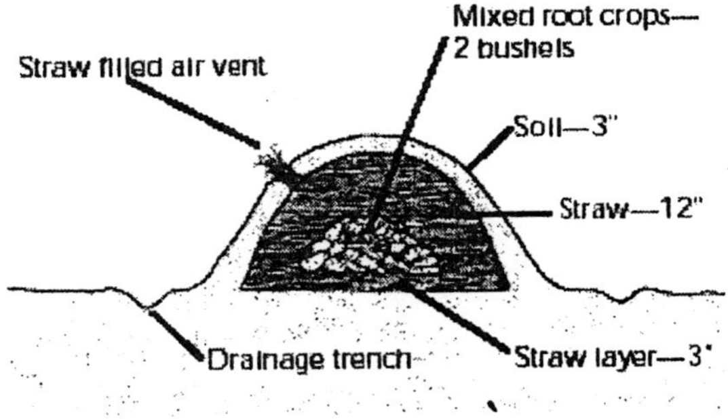

Figure 2. Mound storage.

Making a shallow excavation (from 6 to 10 inches deep) and placing the produce partly below the surface will ensure better frost protection, but it will also increase the danger of excess water. Place mulch over vegetables. A ditch around the storage perimeter will help remove surface water.

Vegetables keep very well in pits and mounds, but once these storage areas are opened all the produce should be removed. After it removed, the produce will keep for 1 or 2 weeks at most. It does not keep as long after removal from storage as will freshly harvested produce. Root crops can be mixed, but should be separated with mulch to prevent cross-transfer of odors.

Buried containers

A 20-gallon trash can, buried in the ground, makes a convenient and economical storage for many kinds of vegetables. Metal cans are more rodent proof than plastic. (fig. 3.)

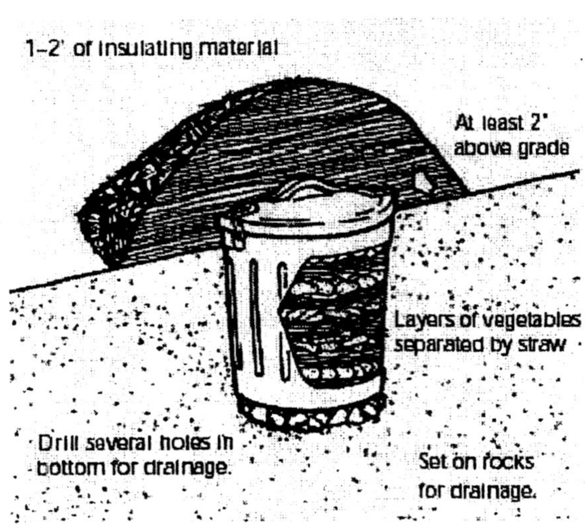

Figure 3. Buried trash can storage.

Several holes should be made in the bottom to facilitate drainage. The container must be free of substances that might give off-flavour to the produce. Never use drums or containers that might have held pesticides or other chemicals.

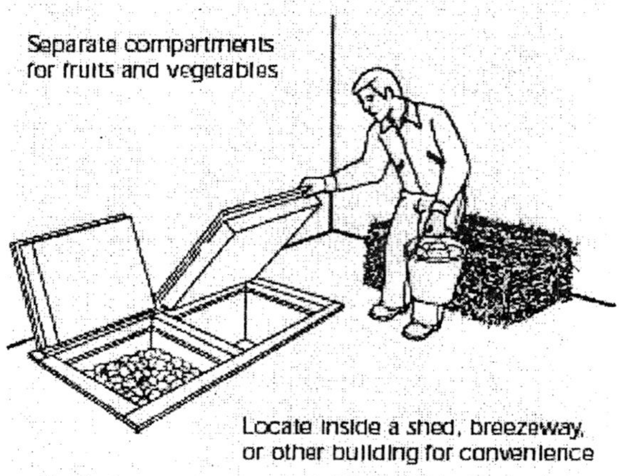

Figure 4. Buried insulated box storage.

An insulated box or other container can be buried in a well-drained area. Buried containers are more easily opened and closed than mounds and trenches. This type of storage could be located in a breezeway, shed, or garage for easier access and greater frost protection. However, if you plan to store food in or near a garage, you must more carefully wrap and protect the produce from car fumes, which produce easily picks up. (fig. 4.)

Indoor Storage Area

There are many areas in dwellings that naturally provide, or can be adapted to provide a variety of temperature and moisture conditions for storage. These include the attic, unheated rooms, the basement, or cellar. (fig. 5) Assess

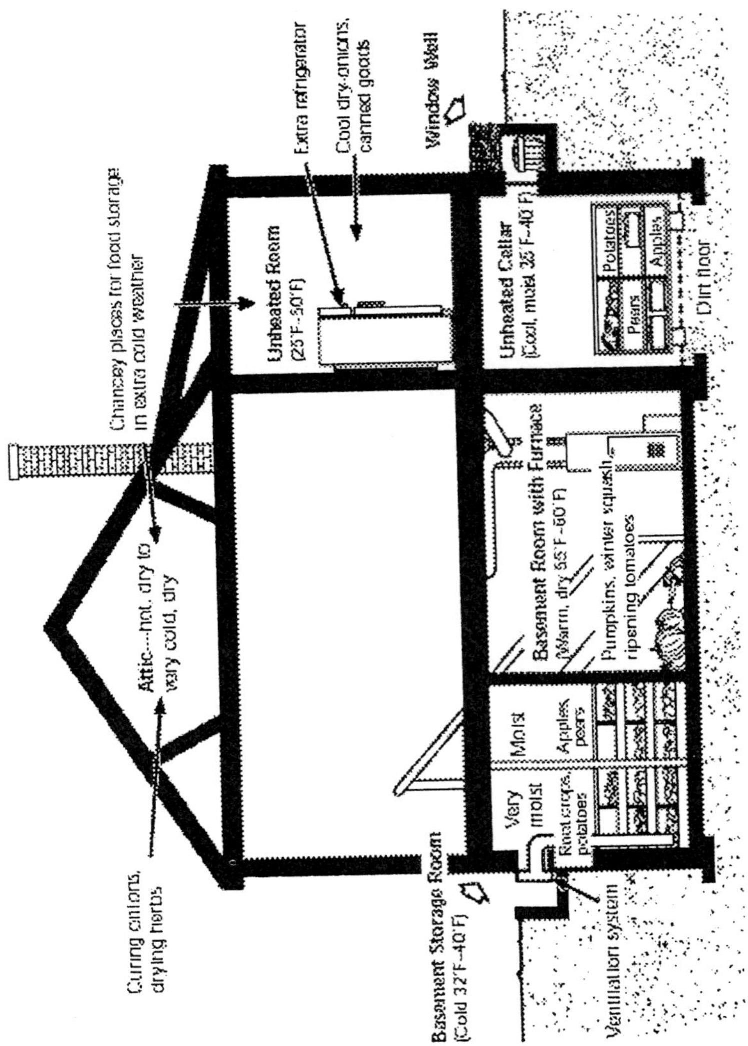

Figure 5. Areas in a house that could be used for storage.

your own storage possibilities. Use a thermometer to monitor temperatures. Any spot that is sufficiently and evenly cool can be adapted for some type of food storage. The relative humidity of these locations will also affect what type of produce can be stored.

Even basement window wells can be converted to small storage areas. They can provide cool, moderately moist conditions and can be used to store some types of apples or a variety of root crops. But be sure to separate the produce properly. Insulate produce with bales of hay or straw. A pantry or unheated room is useful for short-term storage of potatoes and onions, and long-term storage of spices, vegetable oils, nuts, and commercially canned goods.

Low storage temperatures extend the shelf life of dried foods, such as dried beans, herbs, dried fruits and vegetables, and the life of other products, such as coffee, flour, rice, pasta, and cereals. Use containers with tight-fitting lids to keep humidity low and rodents and insects out. Guard against freezing when storing in unheated areas. A warm storage area, such as an attic, can be a good environment in the fall for drying herbs, beans, walnuts, or hickory nuts.

Basement storage

A well-ventilated basement with central heating is generally dry and has a temperature range of 50° to 60°. It may be used for ripening tomatoes and for short-term storage of pumpkins, winter squash, potatoes, sweet potatoes, and onions. If you will be storing only a small amount of produce, an old refrigerator placed in one corner of the basement may be an excellent investment. It can be used for storing head lettuce, cauliflower or cabbage, or apples.

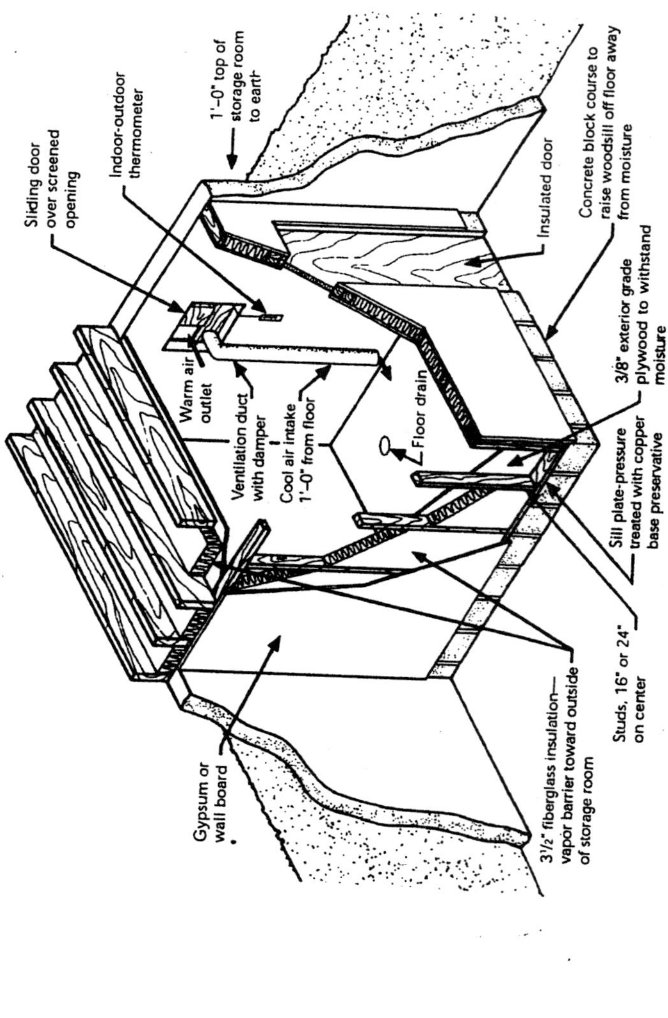

Figure 6. Basement storage.

For long-term storage over winter you will need to partition off a room and insulate it to control temperature and moisture. Build a storage room on the north or east side of the basement, if practicable, and make sure there are no heating ducts or pipes running through it. You will need at least one window for cooling and ventilating the room. Two or more windows are desirable, particularly if the room is divided for separate storage of fruits and vegetables.

A room 4 x 4 x 6 feet is large enough for most households. An area 6 x 7 x 7 feet will hold approximately 30 bushels of produce—5 bushels per month for 6 months. An area 8 x 10 x 7 feet will hold approximately 60 bushels of produce—10 bushels per month for 6 months. (fig. 6) To deter rodents, carefully install sheathings or screens on the partition walls so there are no openings at the floor or ceiling. Young mice can crawl through a quarter-inch hole.

Storage Area Management

Temperature management

Proper temperatures are critical to long-term storage of vegetables and fruits. Improper temperatures, whether too cool or too warm, can cause rapid deterioration of produce quality. Two reliable thermometers (preferably that record minimum and maximum temperatures) should be used to observe and regulate storage temperatures carefully. Place one thermometer in the coldest location of the basement or cellar, and place the other thermometer outdoors. Regulate storage temperature by opening and closing doors, windows, or other openings used as ventilators.

Outdoor temperatures well below 32° are necessary to cool storage air to 32° and to maintain that temperature.

Once cooled to 32°, the temperature will rise again if ventilators are closed, even when outdoor temperature is about 25°. Close ventilators tightly whenever the outdoor temperature is higher than the storage temperature. Watch indoor and outdoor temperatures closely. In most regions, daily adjustment of ventilators is usually necessary to maintain desired storage temperatures.

The danger of overventilating during subfreezing weather is that stored products will freeze if you are not careful. For example, in a study made of an insulated storage cellar partly above ground, full ventilation both day and night was necessary to maintain a temperature of 32° if outdoor temperatures ranged between 18° and 30° during the day and dipped to 10° at night. If minimum temperature at night, with normal air movement, was 8° for 5 or 6 hours, the storage temperature dropped to 30°. During a night of high wind, however, a minimum of 12° cooled the cellar to 30°. Keep a pail of water in the storage area in very cold weather; water will freeze before the crops. However, very low temperatures can cause chilling injury to some produce, like potatoes, and leave them unusable or increase decay.

Managing humidity level

The proper humidity level is also very important to long-term storage and maintenance of good quality produce. This is generally measured as relative humidity, which is the percentage of moisture in the atmosphere at a given time as related to the maximum amount (100%) that could be retained at that temperature. A humidity level of 95% is almost rainfall and rather difficult to achieve indoors. A relative humidity of 90%–5% is very moist and good for storage of potatoes and other root crops.

A relative humidity of 60%—5% is dry and good for storage of pumpkins and other squash. Two ways of maintaining proper humidity are first, to use water to raise the humidity of the storage air, and secondly, to use ventilated polyethylene bags and box liners. Moisture can be added to storage air by sprinkling the floor frequently, by placing large pans of water under fresh-air intake vents, by covering the floor with wet materials such as straw or odorless sawdust, or by a combination of these methods. However, these strategies will not prevent shrivelling of root crops.

The easiest and most effective way to control moisture loss in root crops and in certain other crops is to put them in polyethylene bags or box liners. Cut many ¼- to 3/8-inch holes in the sides of the bags or liners to permit ventilation. Tie the bags and fold over the tops of box liners, but do not seal them. Excessive humidity is conducive to mold and decay organisms, if water droplets are allowed to form on the surface of the product. Plastic bags and liners used for moisture retention should always be perforated at regular intervals to permit air circulation and prevent condensation.

Sanitation

At least once a year, remove all containers from the storage area and clean and air them in the sun. The room itself should also be thoroughly aired, cleaned and washed down with a disinfectant, such as diluted chlorine bleach (1 cup bleach mixed with 9 cups water), to kill off any molds or bacteria that could lie dormant and ruin future crops.

Managing Vegetables and Fruits

Vegetables and fruits that are to be stored should be

handled carefully to prevent damage. Give special attention to containers that you use for harvesting and storing. Use containers that have smooth inner surfaces. Protruding wire staples in baskets and hampers are particularly damaging. Lightweight tub buckets and plastic-coated stave baskets (egg baskets) are good containers for harvesting. If the soil is sandy, rinse the containers frequently to reduce skin breaks.

Standard apple boxes and lug boxes used for shipping tomatoes, grapes, and nectarines are good storage containers. Slatted crates can be made easily from wooden melon crates. Discard all crushed, cut, or decaying vegetables and fruits. If damaged or infected garden products are placed in storage, serious losses from decay are likely to occur. See that vegetables and fruits have as little field heat as possible when you put them in storage. Harvest in early morning, or let the crops cool outdoors overnight before storing them.

Harvest during dry weather, not too soon after a rain, and allow the surface of the product to dry, since wet produce is very much more susceptible to disease. Handle carefully when harvesting and preparing for storage in order to keep bruising to a minimum. Waxing vegetables for home storage is not recommended, although wax has been used for several years on certain fresh vegetables and fruits to improve their sales appearance and to reduce moisture loss.

Separating Fruits and Vegetables

If a large quantity of fruits and vegetables is going to be stored, it is advisable to separate the storage areas or add a central partition in a basement storage area to make two storage areas. Ideally, each area should have its own ventilation system. At least, store fruits and vegetables as far away from each other as possible. Wrapping fruits individually also helps to prevent cross-transfer of odors. Do not store apples with potatoes or carrots.

Ethylene gas produced by apples can cause potatoes to sprout and carrots to become bitter. Potatoes cause apples to take on a musty flavour. Cabbage and turnips can give their odors to celery, pears, and apples. Cabbage, kale, rutabagas, turnips, and winter radishes give off strong odors that could spread through a house, and, therefore, should be stored in outdoor storage areas only.

7

Quality in Fruits and Vegetables

There is a world tendency towards a greater consumption of fruits and vegetables due to a growing concern for a more balanced diet, with a lower proportion of carbohydrates, fats and oils and with a higher proportion of dietary fiber, vitamins, and minerals. Another aspect that deserves attention is the tendency towards simplification in the task of preparing daily meals. In the United States, until the 60s, the preparation of lunch or dinner required about 2 hours and was planned in advance.

Nowadays, meals are prepared in less than one hour and the menu to be served at dinner begins to be defined after 4 p.m. The expanding incorporation of processed fruits and vegetables and other ready-made foods are partly responsible for this reduction in the time dedicated to cooking. Probably, the most significant fact that encourages this tendency is woman's increasing incorporation in full-time work that reduces her time to buy and to prepare foods but giving her more capacity to spend money.

Also influencing the consumption patterns is the increasing market segmentation through the expansion in shapes, colours, flavours, ways of preparation, and/or packaging in which a product is presented. Among others, tomatoes are an example of it, since today they can be purchased in at least 4 different types: conventional or "beef tomato", "extended shelf life", "cherry", and processing types sold fresh, all of them in different sizes, packages and in some cases, colour.

There is also an increasing supply of exotic or non-conventional fruits and vegetables, which together with the previous point, notably expands the purchase options. For example, in 1981, in a well-supplied supermarket of the USA, there were 133 options of different fruits and vegetables, but they increased to 282 in 1993 and to 340 in 1995. Without reaching these levels, the same tendency is observed in the different countries of Latin America and the Caribbean.

Lastly, there is a growing demand for higher quality, external as well as internal quality. External aspects (presentation, appearance, uniformity, ripeness, and freshness) are the main components in the decision to purchase, which is usually taken when the consumer sees the product exhibited at the sales point (Figure 1). This is particularly important in the self-service systems where the product must "self-sell" and if it is not chosen, represents a loss for the retailer. Internal quality (flavour, aroma, texture, nutritional value, and absence of biotic and non-biotic contaminants) is linked to aspects not generally perceived externally, but are equally important to many consumers.

In terms of the service or satisfaction that it produces to consumers, we could also define it as the "degree of fulfilment of a number of conditions that determine its acceptance by the consumer". Here, a subjective aspect is introduced, since different consumers will judge the same product according to their personal preferences.

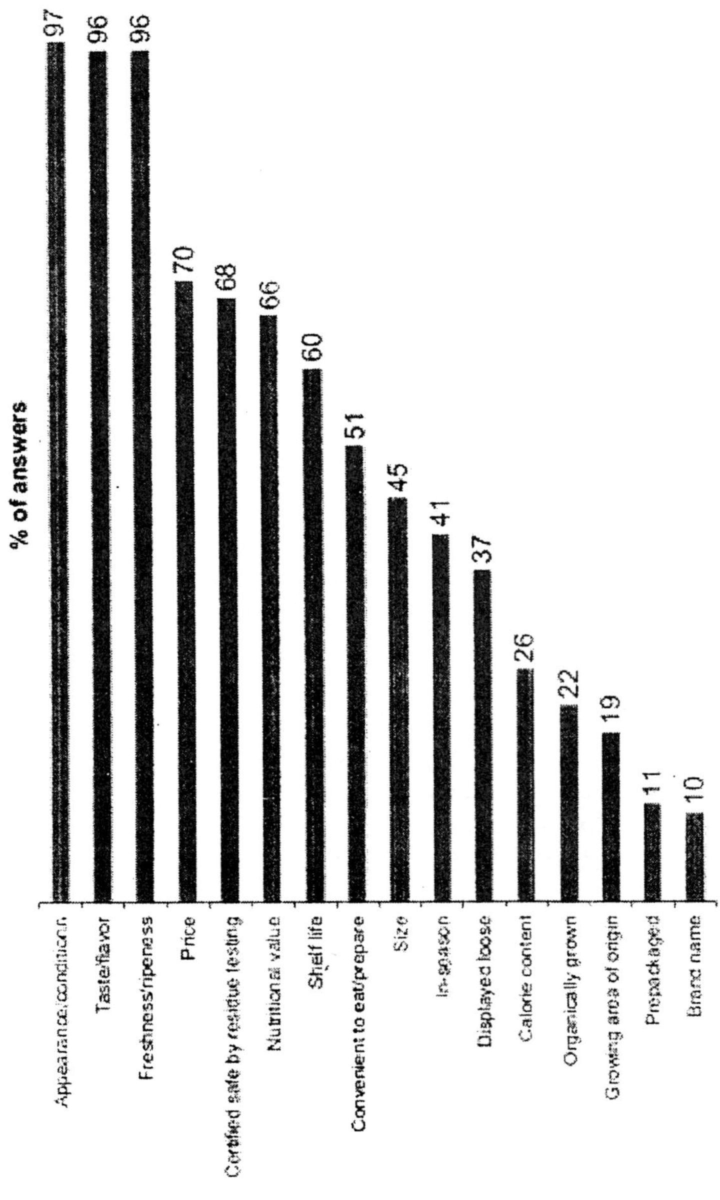

Figure 1. Percentage of answers evaluating these visible aspects as extremely or very important

The destination or use can also determine different criteria for judging quality within the same crop. For example, the tomato for fresh consumption is valued essentially by its uniformity, ripeness, and absence of defects, while colour, viscosity, and industrial yield as raw material define the quality for ketchup tomatoes. It is common to use additional words to define the quality to the specific use, such as "industrial quality", "nutritional quality", "export quality", "edible quality", etc.

Quality: A Complex Perception

Quality is a complex perception of many attributes that are simultaneously evaluated by the consumer either objective or subjectively. The brain processes the information received by sight, smell, and touch and instantly compares or associates it with past experiences or with textures, aromas, and flavours stored in its memory. For example, just by looking at the colour, the consumer knows that a fruit is unripe and that it does not have good flavour, texture or aroma. If colour is not enough to evaluate ripeness, he/she uses the hands to judge firmness or other perceptible characteristics. The aroma is a less used parameter except in those cases where it is directly associated to ripeness like in melon or pineapple. This comparative process does not take place when the consumer faces, for the first time, an exotic fruit whose characteristics are unknown.

Fruits and vegetables are consumed mainly for their nutritive value as well as by the variety of shapes, colours, and flavours that make them attractive for food preparation. When they are consumed raw or with very little preparation, the consumer's main concern is that they must be free of biotic or non-biotic contaminants that may affect health.

Elements of Quality

Appearance

Appearance is the first impression that the consumer receives and the most important component of the acceptance and eventually of the purchase decision. Different studies indicate that almost 40% of the consumers decide what to buy inside the supermarket. Shape is one of the subcomponents more easily perceived, although in general, it is not a decisive aspect of quality, except in case of deformations or morphological defects. In some cases, shape is a ripeness index and therefore an indication of flavour. This is the case of the "full cheek" in mango or the "finger" angularity in bananas.

In those species where the inflorescence is the marketable organ such as broccoli or cauliflower or those that form "heads" like lettuce, cabbage, endive, etc. the compactness is the most relevant feature. In general, it is not associated to their organoleptic characteristics but rather is an indicator of the degree of development at harvest, since open inflorescences indicate that they were picked too late while non-compact "heads" are the consequence of a premature harvest. To a certain extent, it is also an indicator of freshness since compactness decreases with dehydration.

Uniformity is a concept applied to all the components of quality (size, form, colour, ripeness, compactness, etc.). For the consumer it is a relevant feature that indicates that someone that knows the product has already selected and separated it into categories based on the official standards of quality. It is so important that making products uniform is the main activity in preparation for the market.

In many cases, internal or external defects do not affect product excellence, but the consumer rejects them since the absence of defects is one of the main components of appearance and therefore, of the primary decision to purchase. Different causes during growth (climate, irrigation, soil, variety, fertilisation, etc.) can lead to morphological or physiological defects. Some examples of the first ones are "doubles" in cherry, root ramifications in carrots, "catface" tomatoes, "knobby" tubers and "hollow heart" in potatoes, etc.

Tipburn on leafy vegetables and black heart in celery due to calcium deficiencies as well as the internal rots in various species due to boron deficiencies are examples of physiological defects. More serious are those physical or physiological defects that originate during or after preparation for the fresh market and that show up at retail or consumer's level. Within the first ones we have the mechanical damages, bruises or wounds that take place during the handling of the product and that are the entrance doors to most pathogens causing postharvest rots. Chilling injury, ethylene effects as well as sprouting and rooting, are physiological responses to inadequate storage conditions.

Freshness and ripeness are part of the appearance and they have components of their own. They are also indicative of the expected flavour and aroma when products are consumed. "Freshness" is the condition of being fresh or as close to harvest as possible. It is used in vegetables where harvest is the point of maximum organoleptic quality characterised by the greatest turgidity, colour, flavour, and crispness.

"Ripeness" is a concept used in fruits that also refers to the point of maximum eatable quality but that in many cases is reached at the level of sales point or of consumption since, in most commercial operations, fruits are harvested slightly immature. For example, fruits stored in controlled atmospheres reach their eatable quality after leaving the store room, several months after harvest.

Within the parameters for defining freshness and ripeness, colour, both intensity and uniformity, is the external aspect more easily evaluated by the consumer. It is decisive in those products like leafy vegetables or unripe fruits such as cucumber, snapbeans, and others where an intense green is associated with freshness and pale green one or yellowing to senescence. Colour is also an indicator of fruit ripeness and very important in those where no substantial changes take place after harvest (non-climacteric fruits), such as citrus, pepper, eggplant, and cucurbits in general. In fruits that suffer changes after harvest (climacteric) colour is less decisive and basically indicates the degree of ripeness, as for example tomato, pear, banana, etc.

Consumers assign to size a certain importance (Figure 1) and at equivalent quality, intermediate sizes are preferred. In fruits that are naturally large such as pumpkins, watermelons, melons, etc., there is a very defined trend towards sizes that can be consumed by a family (1-2 kg) in a relatively short period (1 week). Size is one of the main indicators of the moment of harvest and in many cases it is directly associated to other aspects of quality such as flavour or texture. Such is the case of zucchinis, peas, haricot beans and miniature vegetables in general where consumers particularly value small sizes.

Gloss enhances the colour of most products, but it is particularly valued in species like apple, pepper, eggplant, tomato, grapes, plums, cherries, etc., to such a point that many of them are waxed and polished to improve their shine. In vegetables, gloss is associated in a certain way to turgidity: a brilliant green is one of the indicators of freshness. It can also be used as a harvest index in eggplants, cucumbers, squash and other fruits that are harvested unripe where the decrease in shine indicates that they have developed too much and have lost part of their characteristics of flavour and texture. On the contrary, in melon, avocado, and other species, it is indicative that it has reached ripeness for harvest.

Different sensations perceived by humans are included within the definition of texture. Thus, firmness is perceived with the hands and, together with the lips, the kind of food surface (hairy, waxy, smooth, rough, etc.), while teeth determine the rigidity of the structure that has been chewed. The tongue and the rest of the mouth cavity detect the type of particles that are crushed by teeth (soft, creamy, dry, juicy, etc.). Also the ears contribute to the sensation of texture, for example, the noises generated when chewing in those species where crispness is an important aspect.

Together with flavour and aroma, texture constitutes the eating quality. An over-ripe tomato, for example, is mainly rejected by its softening and not because important changes in the flavour or aroma have taken place. Although it is decisive for the quality of some fruits and vegetables, in others it has a relative importance. In texture terms, each product is valued differently: either for its firmness (tomato, pepper), the absence of fibers (asparagus, globe artichoke), its softness (banana), juiciness (plums, pears, citrus), crispness (celery, carrot, apple), etc.

Firmness and colour are the main parameters to estimate the degree of ripeness of a fruit since this process initially improves and softens fruit texture, which together with the changes in flavour and colour, bring the fruit to reach its maximum eatable quality. However, as this process continues, over-ripeness takes place, which leads finally to tissue disorganisation and decay of the product. Firmness is used mainly as a harvest index and it is measured with instruments that register the force necessary for a certain deformation or resistance to the penetration of a piston of known dimensions.

Juiciness is the sensation of liquid spilling inside the mouth as tissues are chewed. The juice content of many fruits increases as they ripen in the plant. It is regulated that the minimum content that citrus fruits should have, is: 30% for Navel oranges; 35% for grapefruit and the other oranges; 25% for lemons; 33% for mandarins and 40% for clementines.

Flavour

Flavour is the combination of the sensations perceived by the tongue (taste) and by the nose (aromas). Although those sensations can be perfectly separated one from the other, as the sensitive receptors are so close, simultaneously with the act of bringing near the mouth, of biting, chewing, and tasting, we are perceiving the aromas, particularly those that are liberated with the crushing of tissues. It is also possible, however, that certain external aspects (particularly ripeness) anticipate the flavour that should be expected when consuming the product. The human being has stored in his/her memory an enormous quantity of different tastes and aromas and, if it was eaten previously, is able to recognise them without seeing the product.

In fruits and vegetables, taste is usually expressed in terms of the combination of sweet and sour principles that are an indication of ripeness and eating quality. The content of soluble solids is a good estimate of total sugar content, and many fruits should have a minimum content of solids to be harvested (Table 1). Organic acids (citric, malic, oxalic, tartaric) are the other important components of taste, particularly in their relationship with soluble solids.

As the fruit ripens they tend to diminish and so the relationship with the soluble solids tends to increase. Titratable acidity is the form of expressing acidity. The soluble solids/titratable acidity relationship is a denominated ratio and it is essentially used in citrus where it is a function of the species and of the variety. Its value is 8 for mandarins, Navel oranges, and hybrids, 7 for other type of oranges, and 5.5 for grapefruits.

Astringency (sensation of loss of lubrication in the mouth) and bitter tastes are due to different compounds. They are not frequent and when they exist, usually diminish with ripening. In those cases in which they appear naturally and represent a disadvantage, they have been eliminated through breeding programmes.

There are specific compounds that characterise certain species or a group of them, for example, pungency in the peppers denominated "hot" is basically determined by the capsaicin content and other 4 structurally similar compounds. There are also cases in which enzymes and substrata responsible for the taste are compartmentalised in healthy tissues and they only get in contact by cutting, chewing or crushing.

This is the case of pungency in garlic and onion and also of the taste of raw cucumber. Cooking these vegetables whole prevents these reactions and the resulting taste is different.

Table 1. Recommended minimum soluble solid content at harvest.

Apple	10,5-12,5
Apricot	10
Blueberries	10
Cherry	14-16
Grape	14-17,5
Grapefruit	8
Kiwifruit	14
Mango	8
Mandarin	8
Melon	10-12
Nectarine	10
Orange	8
Papaya	11,5
Peach	10
Pear	13
Persimmon	18
Pineapple	12
Plum	12
Pomegranate	17
Raspberry	8
Strawberry	7
Watermelon	10

There is a correlation between dry matter content and organoleptic characteristics mainly used by the industry. In general, a higher content of solids means higher industrial yield and taste. This is particularly important in dehydrated products. In potato, a higher content of dry matter (measured as specific gravity) is associated to a better cooking quality. For the fresh market, however, dry matter content is not used as an indicator of the time of harvest and/or organoleptic quality, except in the case of avocado where there is a close correlation with the oil content. Depending on the variety considered, avocados with a dry matter content lower than 21-23% should not be marketed.

The aroma of fruits and vegetables is due to the human perception of numerous volatile substances. Refrigerated fruits and vegetables are less aromatic since volatile liberation diminishes with temperature. As well as in the case of taste, many aromas are liberated when tissues lose their integrity.

Nutritive value

From the nutrition point of view, fruits and vegetables are insufficient to satisfy daily nutritional requirements, essentially because of their low content of dry matter. They have a high content of water and are low in carbohydrates (excepting sweet potatoes, potatoes, cassava, and other underground organs), proteins (except for legumes and some crucifers) and lipids (except avocados), but they are, in general, a good source of minerals and vitamins. Different countries have made tables of recommended daily consumption, the best known being probably the U.S.R.D.A. (United States Recommended Daily Allowances) (Table 2). These tables are only for reference and they indicate the capacity of foods to satisfy the daily needs for certain nutrients. The conditions of cultivation, varieties, climate, and preparation affect the actual content of nutrients.

Dietary fiber can be defined as the vegetable portion that cannot be degraded by the enzymes of the human digestive tract although its components are anaerobically metabolised in variable proportions by the colon microflora. The optimum daily ingest for a healthy adult is 25 to 30 grams and the contribution of certain fruits and vegetables to the minimum needs is shown in table 2.

The discovery that certain foods have biologically active compounds, beneficial to health beyond basic nutrition opened a new stage in nutrition science. These compounds or their metabolites that have been denominated "functional", help to prevent diseases like cancer, have a protective effect on cardiovascular problems, are neutralisers of free radicals, reduce cholesterol and hypertension, prevent thrombosis, besides other beneficial effects.

Those foods that contain them are also called "functional" although other names such as "pharmafoods", "nutraceuticals", and others have been proposed. As most of these compounds are of plant origin, many authors call them phytochemicals. Fruits and vegetables are rich in phytochemicals like terpenes (carotenoids in yellow, orange and red fruits and limonoids in citrus), phenols (blue, red and purple colours of cherries, grapes, eggplant, berries, apples and plums), lignans (broccoli), thiols (sulfur compounds present in garlic, onion, leek and other alliums, cabbages and other crucifers) and others.

Table 2. Content of essential nutrients (= good; X = very good; XX = excellent; •= exceed daily requirements).*

SPECIES	Vitamin A	Vitamin C	Thiamin	Riboflavin	Niacin	Calcium	Iron	Fiber
Swiss chard	X	*						
Artichoke		*						XX
Pineapple		X						
Celery		*						X
Blueberry		X						*
Cranberry		*						*
Pea		XX	*					*
Banana		X						
Sweet								

Contd....

Contd....

potato	•	XX		*				*
Broccoli	*	•		*				•
Carambola	*	XX						*
Onion		*						X
Plum		*						
Cauliflower		•						XX
Kohlrabi		XX						
Kale	•	•						
Snapbeans		*						
Cherimoya		XX		*	*			XX
Parsnip		*		*				
Sweet corn		*						
Apricot	XX	*						
Date		*						
Belgian endive	*							
Asparagus		*	*					XX
Spinach	XX	X					*	
Raspberries		XX						XX
Strawberry		•						
Fig						X	*	XX
Mushrooms				*				
Kiwi		•						XX
Lettuce Iceberg							*	
Lettuce Romaine	X	X						
Lime		X						

Contd....

Contd....

Lemon		XX						
Mango	•	XX	*					
Apple		*						X
Cantaloupe ripe	•	•						*
Melon Honeydew	XX						*	
Blackberry		X						X
Turnip		*						
Orange Navel		•						
Orange Valencia		XX						
Nectarine		*						
Okra	*	*						
Avocado		*						XX
Potato	*	XX	*	*	*	*		X
Papaya		•	*		X			
Cucumber								*
Pear		*	X					
Pepper		•						
Grapefruit		XX						
Leek		*						
Radish		X						
Beet (topped)		*						
Cabbage	XX	•						XX
Cabbage								

Contd....

(red)	XX			
Cabbage Savoy	*	X		
Brussels sprouts	*	•		*
Rutabaga		X		
Water-melon	X	XX	X	*
Tomato	*	XX		
Jerusalem artichoke		*		
Carrot	•	*		
Pumpkin	X	*		
Pumpkin acorn		X	*	*
Pumpkin Butternut	•	XX		*
Pumpkin Hubbard	•	*		*

Safety

Fruits and vegetables should be attractive, fresh, nutritive and with a good appearance and presentation. Besides these characteristics, their consumption should not put health at risk. The consumer has no way to detect the presence of dangerous substances on food and he/she depends entirely on the seriousness and responsibility of all the members of the production and distribution chain. Necessarily, he/she has to trust them, in addition to the usual precautions of washing, peeling and/or cooking the product before consuming it.

However, this trust is very volatile and any suspicion about safety has a tremendous impact at consumer level.

Among the most relevant examples it is worth to mention the epidemic of cholera in the 90s in Latin America that reduced the consumption of vegetables in many countries of the region for almost one year. Another example may be the two grapes with dangerous residues detected in the 80s in an entrance port of the USA, which severely affected Chilean exports. Also about that time, the Alar (daminozide, a growth regulator used to control fruit set) scare considerably diminished the consumption of apples in the United States.

Food safety is the absence of substances dangerous for health and particularly in fruits and vegetables, the presence of pesticide residues on the product has been the main concern for consumers. However, there are many other contaminants potentially as dangerous such as the presence of pathogenic microorganisms, mycotoxins, heavy metals, and others.

As fruits and vegetables are consumed fresh and are many times not peeled, all organisms pathogenic to humans which are carried on their surface constitute a potential danger. Bacteria, like Shigella spp., Salmonella spp., Aeromonas spp., Escherichia coli, Listeria monocytogenes and the toxins produced by Clostridium botulinum and others, have been identified as responsible for illnesses associated with the consumption of fruits and vegetables. The Hepatitis A virus has been detected on produce as well as parasites like Entomoeba histolyca, and Giardia lamblia.

Agrochemicals are one of the tools that man has used to satisfy the growing need for food. They are the herbicides, insecticides, fungicides, fumigants, rodenticides, growth regulators, waxes, disinfectants, additives and all other products of a chemical nature used during production or post-harvest handling. Their residues have always been a major concern, although the advances in chemistry and biochemistry, the better understanding of ecology as well as pests and diseases, and the growing use of non-chemical control methods, have made the present world a much safer place. Today's agrochemicals are less toxic and persistent, their degradation products are generally innocuous and better laboratory methods have been developed for their detection. Additionally, there is a greater consciousness about their use, waiting times, doses, etc.

Each country has its own legislation in terms of the Maximum Residue Levels (MRLs) within the framework of the Codex Alimentarius Commission or other international organisations. An MRLor tolerance is the maximum concentration of pesticide residue allowed resulting from its application according to correct agricultural practices.

Agrochemical use should be within the Good Agricultural Practices framework to guarantee maximum safety and to minimise risks to the consumer's health. Specific products should be used to control pests or diseases following the manufacturer's indications, particularly those refering to crops on which they can be used, as well as minimum waiting times between application and harvest.

Other health hazards are the presence of nitrates in leafy vegetables, oxalates in some species and heavy metals accumulation, particularly when sewage is used as fertilisers or organic amendments. Some toxicity may exist in some natural compounds produced by the crop itself or by the fungi that colonise its surface like the micotoxins.

Producing Quality Product

Producing a quality product begins well before planting

the seed. Soil selection and preparation, its fertility and irrigation aptitude, weed control and crop rotations, variety selection and other decisions have an influence on the quality of the product. In the same way, quality is affected by the climatic conditions during the growing period, as well as irrigation, fertilisations, control of pest and diseases and other cultural practices. Harvest is the end of cultivation and the beginning of post-harvest actions during which preparation for the market, distribution, and sale take place.

Fruits and vegetables are highly perishable products and before being detached from the mother plant all demand water and nutrients. Once harvested, however, they depend on their reserves to continue living. Respiration, transpiration and the continuous changes taking place determine the internal and external quality. Deterioration rates depend on the type of product, growing conditions and other factors, but mainly on the conditions in which the produce is maintained after harvest such as temperature, relative humidity, movement and composition of the air, etc. Post harvest changes can only be delayed within certain limits and thus preparation for the fresh market should be quick and efficiently performed to avoid quality losses.

Post-harvest losses due to microorganisms can be severe, particularly in warm climates with a high relative humidity. Rotting produce contaminates the rest and under these conditions ethylene production is stimulated accelerating the rate of deterioration.

Most of the fungi and bacteria that attack fruits and vegetables after harvest, are weak pathogens and they mainly invade tissues through wounds. Injuries produced during handling provide numerous entrance routes to these pathogens although some of them are able to invade healthy products. Unripe fruits are usually more resistant to pathogen attack. It is also possible that infection occurring at the immature stage shows up later, when the natural defences are weakened by the ripening process.

A good disease control programme at field level reduces the source of inoculum and the risks of infections after harvest facilitating the control of post harvest diseases. Also, careful handling during harvest and packing operations reduces the physical damage that facilitates the establishment of microorganisms. Controlling temperatures to which produce is exposed is one of the main tools to control post harvest diseases since it diminishes the metabolic activity of the microorganisms and, by reducing the rate of the ripening process, the natural defences of the product are kept high. Controlling the relative humidity, particularly to avoid the condensation of water on the product, as well as controlled atmospheres is also useful in the control of post-harvest diseases.

Total Quality in Fruits and Vegetables

The concept of quality as a way to differentiate products has been recognised for years. As local or regional trade internationalises, quality consolidates as the main competitive tool for excellence, reinforcing the need to establish standards to separate quality into categories or degrees, as well as to define the limits of allowed defects. Nowadays, domestic and international trade of fruits and vegetables is regulated by quality standards in most countries, providing a common language among the different participants of the production-commercialisation-consumption chain.

Standards are also the legal framework to settle commercial disputes and are useful as a basis for reporting on market prices as prices only can be compared between the same quality category. The quality system established by the standards is known as "Inspection for quality" (Table 3) where representative samples at the final stage of preparation for the market should fulfill the specified limits and their tolerances. Although it is easy to apply, it has, at least, two big disadvantages: firstly, they are not totally adapted to highly perishable products where quality varies continually. Secondly, its application does not improve the quality of the product, it only separates in degrees the quality that comes from the field.

At the same time that quality standards developed and were applied, new ideas began to be conceived by industry. Firstly, it became evident that a systematic and preventive approach was much more effective and economic to improve quality than the elimination of the faulty units at the end of the line, when the production and packaging costs had already happened. Secondly, it became clear that the quality concept extends beyond the product itself, since it is affected by the systems and procedures involved in the production and preparation for the market. Finally, the consumer's opinion begins to be more and more important. It is no longer sufficient for a product to be technically perfect and produced in an economically profitable way, it is also necessary to satisfy the consumers'expectations of quality.

The application of statistics to control the variability of the different units in the production lines gave birth to the system called "Quality control" or "Statistical Quality Control", which was adopted by most manufacturing companies in the first half of the 20th century. This method or system essentially provides the analytic tools for monitoring the production process and for taking measures when variability exceeds certain limits considered as normal. Its application improves the quality of the process contributing greatly to improve the quality of the product. They are tools that can be applied at the fruit and vegetables packinghouse level.

This system was transferred to Japan after World War II where it evolved into what today is known as "Total Quality Management" or simply "Total Quality". Total quality is today the most complete conceptual framework to assure quality to which each person or activity within the production process is committed, aiming at zero defects and customer's complete satisfaction, even going beyond his/her expectations. At the same time that TQM was developed, the concept of "Quality Assurance" was coined in Europe.

Its scope is slightly narrower than TQM, but a lot easier to implement and probably better adapted to fruit and vegetables. It is defined as all those planned and systematised actions necessary to guarantee that the product or service will satisfy the requirements of quality. It usually requires the fulfilment of certain rules, protocols, or standards developed specifically and with a certification by an independent company authorised to grant it. The ISO system is probably the best known and within it the series 9000.

Table 3. Comparison of the main quality systems

Aspects	Quality inspection	Quality assurance	Total quality
System	Reactive	Preventive	Preventive
Quality is	A control procedure at the end of the process	The objective of an explicit policy	A philosophy
Application of regulations	Only the mandatory ones	Mandatory + voluntary ones	Mandatory + voluntary of

Contd....

Contd....	(Standards)	as ISO, HACCP	own design
Quality is based on	The final product	The organization	Human resources
Quality control is performed by	A quality laboratory	Quality management level	All
Documentation on processes and methods	No	Yes	Yes
Internal auditing	No	Yes	Yes
Certification of conformity	No	Yes	Not necessary

It is also appropriate to mention the HACCP (Hazard Analysis Critical Control Points) system, designed specifically to guarantee that food is not exposed to any type of contamination that could put health at risk. Today, this method is recognised internationally as the logical and scientific tool for all food quality systems. It is also preventive in nature and the key element is the identification of the critical points, within the process, where quality should be controlled to prevent, eliminate or reduce to acceptable levels all possible safety risks.

The HACCP system is required today in the USA and other countries to import meat, fish, eggs, and other foods. Up to now, it is not required for fruits and vegetables, although different export countries are already implementing it to assure a superior quality of their products. The logic of the HACCP can be applied to the detection of other defects of quality.

Although all these systems have their origin in industry, their application extends to other sectors. Agriculture, and particularly the production of fruits and vegetables, is now incorporating many of the methods and ideas conceived by the industrial sector because the basic principles are not only applicable but also recommendable for highly perishable products where quality deteriorates quickly.

Several export companies have implemented the HACCP together with the ISO 9002 certification, which guarantees food safety within a system of quality assurance. Akey concept is that quality systems are not mutually exclusive but rather they overlap widening the application approach, extending beyond the product itself and embracing the preparation process, inputs, suppliers and intermediaries, besides the incorporation of the feedback from the client or consumer for its continuous improvement.

The basic principles of total quality can be summarised in the following way:

- The consumer is always first
- Each operation is part of a process
- Quality improvement never ends
- Quality is made, not controlled
- Prevention of quality problems is made through planning.
- The desired product should be obtained at the desired moment. Post harvest handling should be appropriate to reach the desired market under the desired conditions.

8

Hygiene and Sanitation

Food safety has always been a main concern and its origins can be traced back to very early times. Ensuring food safety is based on the principle of sound science and its proper application. This applies also to fruits and vegetables.

Microbiological Risk in the Production and Distribution

Products go through different stages of operations after harvest. This provides many opportunities for contamination besides those which naturally occur in the field. Consumers strongly reject foreign materials on products or inside packages. These include for, example, dirt, animal feces, grease or lubricating oils, human hairs, insects, plant debris, etc. However, because this is usually due to insufficient care in handling, they are relatively easy to detect and to eliminate.

A more serious problem is the presence of human pathogens on produce. These may not be visible or detected because of changes in appearance, flavour, colour or other external characteristics. It has been shown that specific pathogens are able to survive on produce sufficiently long to constitute a threat. In fact, many cases of illness related to consumption of produce have been reported (Table 1).

Three types of organisms can be transported on fruits and vegetables, which may constitute a risk to human health: virus (hepatitis A, for example), bacteria (*Salmonella* spp., *Escherichia coli, Shigella* spp., and others) and parasites (*Giardia* spp., for example). Mycotoxins and fungi do not usually constitute a problem. This is because fungi development is usually detected and eliminated well before the formation of mycotoxins. In most cases, bacteria is responsible for illnesses related to the consumption of fruits and vegetables.

Table 1. Isolated pathogens in fruits and vegetables and reported illnesses

Aeromonas spp.	Alfalfa sprouts, asparagus, broccoli, cauliflower, lettuce, pepper
Bacillus cereus	Sprouts
Escherichia coli O157:H7	Cabbage, celery, cilantro, lettuce(*), pineapple, apple cider(*), alfalfa sprouts(*)
Listeria monocytogenes	Bean sprouts, cabbage, cucumber, shredded cabbage(*), potato, radish, mushrooms(*), salads(*), tomatoes and other vegetables
Salmonella spp.	Artichoke, bean sprouts(*), tomato(*), alfalfa sprouts(*), apple cider(*), cauliflower, celery, eggplant, Belgian endives, pepper, cantaloupe(*), watermelon(*), lettuce, radish and several vegetables
Clostridium botulinum	Shredded cabbage(*)
Shigella spp.	Parsley, leafy vegetables, shredded lettuce(*)
Cryptosporidium spp.	Apple cider(*)
Cyclospora spp.	Raspberry(*), basil(*), lettuce(*)
Hepatitis A	Lettuce(*), strawberry(*), frozen strawberry(*)

(*) Reported illnesses.

Produce can become contaminated through mechanisms that are complex (Figure 1). The best strategy to obtain a safe product is to prevent contamination at various points throughout the production and distribution chain.

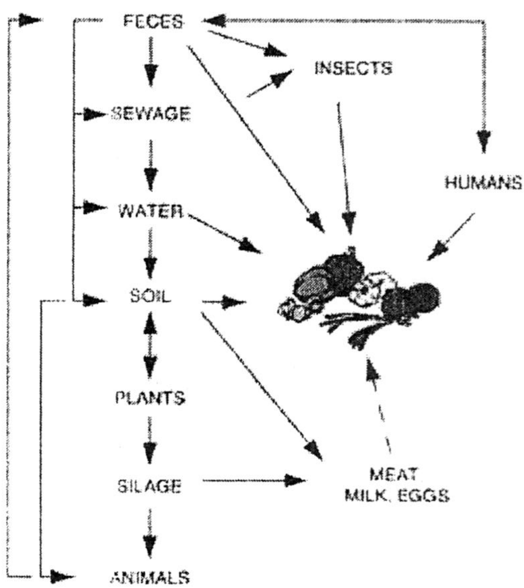

Figure 1. Mechanisms by which fruit and vegetable can become contaminated with pathogenic microorganisms

This includes undertaking certain sanitary treatments and maintaining produce under conditions (mainly temperature) which do not favour the development of micro-organisms. This approach is known as the "systems approach". Every step of the process is a part of an integrated system.

Records and/or documentation of all activities and treatments are required so that a tracing scheme is established. In this way, it is possible to identify areas of weaknesses within the system and to take corrective measures. Following strict written procedures for good agricultural practices (GAP) and/or good manufacturing practices (GMP) are critical factors for the implementation of such a system. This should be combined with such systems as Hazard Analysis Critical Control Point (HACCP) analysis in order to identify the critical points at which known safety food hazards must be controlled.

Before Harvest

Some human pathogens are naturally present in the environment. However, fecal deposits(human, animal, or wild animals) are the main source of contamination of produce. Entry is mainly through irrigation or washing water. Microorganisms in surface water (rivers, lakes, etc.) may come from upstream dumping of untreated municipal wastes. Underground water may also be contaminated from septic tanks leaching through soil into aquifers. If only contaminated water is available, underground drip irrigation is the only irrigation system recommended to avoid contamination of above-ground edible plants (Table 2).

The main causes of contamination include: animal manure or sewage waste used as organic fertilisers or the presence of animals in production areas. Manure should be composted aerobically to reach 60-80°C for a minimum of 15 days. Composting of static piles and earthworms do not guarantee that microorganisms have been inactivated. Wastewater and municipal wastes should only be used if effective disinfecting systems are available.

Table 2. Potential risks of microbial contamination and recommended preventive measures.

Production step	Risks	Prevention
Production field	Animal fecal contamination	Avoid animal access, either wild, production or even pets.
Fertilising	Pathogens in organic fertilisers	Use inorganic fertilisers. Proper composting

Contd....

Contd....

Irrigation	Pathogens in water	Underground drip irrigation Check microorganisms in water
Harvest	Fecal contamination	Personal hygiene. Portable bathrooms. Risk awareness
	Pathogens in containers and tools	Use plastic bins. Cleaning and disinfecting tools and containes.
Packhouse	Fecal contamination	Personal hygiene. Sanitary facilities. Avoid animal entrance. Eliminate places may harbor rodents.
	Contaminated water	Alternative methods for precooling. Use potable water. Filtration and chlorination of recirculated water. Multiple washing
Storage and transportation	Development of microorganisms on produce	Adequate temperature and relative humidity. Watch conditions inside packaging. Cleaning and disinfection of facilities. Avoid repackaging. Personal hygiene. Do not store or transport with other fresh products. Use new packing materials
Sale	Product contamination	Personal hygiene. Avoid animal access. Sell whole units. Cleaning and disinfection of facilities.Discard garbage daily.

The production and harvest of fruits and vegetables tend to rely heavily on human labour. Other sources of contamination include the hygiene conditions of field workers. First, production fields are usually a long distance from bathrooms, together with other sanitary facilities for the hygiene of personnel, Second, hired teams of migrant labour temporarily live in the fields. Here, conditions and sanitary practices are considered unacceptable. In addition to providing portable toilets, staff need to understand the importance of proper hygiene practices for food safety.

The type of product also has an influence: in vegetables which are characterised by low acidity in tissues, bacteria tends to dominate, while fruits are mainly colonised by fungi. Crops growing close to the ground like strawberry and leafy vegetables in general, are more susceptible to contamination by water, soil, or animals when compared to tree crops. Finally, some chemical constituents of tissues such as organic acids, essential oils, pigments, phytoalexins etc, have antagonistic effects and provide some form of protection against the development of micro-organisms.

As with other handling operations, numerous opportunities exist for contamination during harvest. Wounds and bruises may exude latex and other plant liquids from tissues and provide the substrate for the growth of microorganisms transmitted by hands, tools, clothes, water, or containers. Contamination at any point throughout the distribution chain can be exacerbated by the conditions to which the produce is exposed. Temperature is the most important factor to be considered.

Preparation of Product for Market

In packinghouses or processing plants, people who are ill

or have open wounds should be forbidden from making contact with the product. Workers should use hairnets and clean outfits when handling the product.

Street clothes must be left outside the working area and eating or drinking in the pack-house should be forbidden. Workers should wash their hands daily at the beginning of operations and every time they return to work, particularly after having used restrooms.

However, the main source of contamination in terms of product preparation for market, is probably water. Water is essential for packinghouse operations either for washing the product, containers, and facilities or for dumping, hydrocooling. Other uses include personal hygiene or as a medium for waxes, chemicals, etc.

Water disinfection

Water impurities are frequently concerned with: the suspension of materials, microorganisms, organic matter, off-colour, off-odour as well as minerals and dissolved gases. Municipal water is filtered and treated (normally with low chlorine concentrations) to ensure that it meets the chemical and microbiological requirements for food safety and contact with food. Alternative sources of water must be filtered and sanitised.

Sanitation is necessary to avoid the spread and contamination to other units, even with the use of municipal water. Different methods exist for water disinfection. These include chemical, thermal, ultrasonic waves or irradiation. In postharvest operations, chlorine and its derivatives are the cheapest and most widely used substances to destroy bacteria and fungi in water as well as on the surface of fruits.

Chlorine is a gas with a strong and penetrating odour and is extremely reactive chemically. At post harvest level, it is used mainly in three different forms: as pressurised gas from metal cylinders, as calcium hypochlorite (solid) or liquid as sodium hypochlorite, commonly known as "bleach" for household whitening and sanitising. Chlorine gas is difficult and dangerous to handle and is normally limited to large operations such as municipal water treatment. Calcium hypochlorite is widely used in concentrations of 65% but is difficult to dissolve in cold water. Sodium hypochlorite is more expensive than the other two formulations in terms of chlorine concentration (5 to 15%). However, its easy dosage makes it convenient for small size operations.

In aqueous solution, chlorine exists as hypochlorous acid, hypochlorite ion, or as a variation of both, depending on the pH solution: the former predominates in acidic solutions while the latter in alkaline solutions. Germicidal action of hypochlorous acid is about 50-80 times higher than hypochlorite. In order to maximise the effect on microorganisms, the pH solution should have a range of between 6.5 and 7.5. Below this range, the hypochlorous form is extremely unstable and tends to escape as a gas resulting in irritation and discomfort to workers.

It also becomes extremely corrosive for the equipment. On the other hand, its effectiveness as a sanitiser is significantly reduced above 7.5. To keep pH values within permitted ranges, vinegar could be used to acidify while sodium hydroxide could be used to alkalinise. Maintenance kits for swimming pools could be used to monitor pH. Chlorine as a gas lowers the pH of the solution while as hypochlorite, either of calcium or sodium, this will be increased.

Concentration of active chlorine is expressed in parts per million (ppm). Concentrations of active chlorine in the range of 0,2 to 5 ppm are able to kill most bacteria and fungi present in water. However, in commercial operations higher concentrations are used (100-200 ppm) for washing and hydrocooling. Aliter of household bleach (80 g active chlorine/dm^3) in 400 liters of water represents about 200 ppm, and in 800 and 1600 liters, about 100 and 50 ppm, respectively. It is convenient to start daily operations with: low concentrations (100-150 ppm); to increase the amount of chlorine in solution as water becomes contaminated with dirt and plant debris and an increase in microorganisms.

Quick exposure (around 3-5 minutes) is adequate for the purposes of disinfection. However, in addition to pH level and impurities, the temperature of the solution is also important. This is because low temperatures can reduce activity. Another important factor is the extent to which microorganisms can be developed. This is because their spores are from 10 to 1000 times more difficult to kill compared to their vegetative state.

The use of chlorine in fruits and vegetables is banned in some countries. This is because it is possible for it to react with organic matter leading to the formation of chlorate compounds and trihalomethans. These are suspected of being carcinogenic. As a result of this, the industry is investigating alternative sources of sanitisers.

Ozone is a gas with a strong oxidizing action in concentrations of 0,5-2 ppm. It has been approved for water sanitation. However, it is difficult to apply. This is because reliable methods for monitoring the level of concentrations do not currently exist. In addition to this, it is only effective within a reduced pH range of (6-8) and must be generated in the same place of application. It is dangerous to humans in concentrations higher than 4 ppm and may cause damage to some plant tissues. In spite of these limitations, it is probably the most promising compound to replace chlorine. Ultraviolet light in wavelengths of 250-275 nm can also be used. It is unaffected by water temperature or pH. However, water must be filtered as turbidity reduces its efficacy.

Water management is also important. This is because several washes are required. This is more effective than a single wash. The following is an outline of a procedure for good cleaning practices: First, undertake an initial wash to eliminate dirt and plant debris. Second, wash with chlorinated water and finally rinse with plain water. Brushing or water agitation increase washing efficiency. Water recirculation should be undertaken in reverse to the flow of the product, i.e. rinsing waters can be reused for the initial washing. Hydro cooling is one of the most efficient systems for precooling. It is, however, also one of the most risky in terms of microbial contamination. This can be caused by water infiltration inside the fruits. As a result of this, it is important to consider alternative precooling methods such as forced air.

Plant hygiene

Industrial facilities follow a system of strict hygiene measures. However, when produce is prepared for the fresh market, limited attention is normally paid to hygiene facilities. This is particularly the case when low cost materials have been used in the construction of the packing shed.

Although other factors need to be taken into account prior to layout and organisation, it is important that the packing shed be designed to allow for thorough cleaning procedures. The reception area should be kept separate from the delivery area. Similarly, "clean areas" or areas where product is prepared, should be kept separate from other areas or where product is handled as receivables from the field. There should be a clean area where workers can take breaks, change clothes, and take care of personal hygiene. For example, the availability of hot water, showers, and clean restrooms within a clean and comfortable environment.

While it is a requirement that dust and other impurities are eliminated, liquid sanitisers should be used to disinfect the facility and equipment, particularly those in contact with produce. Chlorine based sanitisers are the most widely used disinfectants. However, the choice depends on type of water, pH, cost and the type of equipment. Iodine based disinfectants (iodophors) are less corrosive to metals than chlorine. They are unaffected by organic matter but their effectiveness within the pH range is quite narrow (2,5-3,5). In addition to this, they may stain surfaces. Quaternary ammonium compounds are widely used for disinfecting floors, walls, and aluminum equipment. While effective over a wide pH range, they are not affected by organic matter and are non-corrosive. However, they are expensive and leave residues on surfaces. There are other sanitisers available on the market that can be used in food plants.

Animals of all types, including mammals, birds, reptiles, and insects can spread microorganisms with their droppings. Their entry into packing areas should be forbidden, including pets. It is necessary to seal crevices and keep doors, windows, and air inlets closed or fitted with insect-proof screens. It is also important to have insect and rodent control programs, together with approved pesticides, traps, and baits. Facilities and the surrounding environment must be kept clean and tidy to prevent the nurturing and harbouring of insects, rodents, reptiles and other animals. Garbage and waste materials should be removed daily.

Storage and transport

In terms of sanitation, there are two potential sources of risk: contamination by human pathogens and their growth and development due to the favourable conditions created because of the packaging environment or storage area. Factors highlighted earlier about hygiene of personnel and facilities are also applicable here. Two additional recommendations include: the use of new containers and not undertaking repackaging. Other considerations to prevent cross contamination include not storing or transporting fruits and vegetables with other fresh food items.

The best strategy to prevent growth and development of human pathogens is to keep product at recommended storage conditions, particularly temperature. Microorganisms can be divided into three main categories according to their level of adaptation to temperature. These are: psychrotrophic - the ability to grow under refrigerated conditions, although the optimum ambient temperature is (20-30°C); b) mesophilic - those that develop best at ambient temperature (20-40°C) but not under refrigeration and c) ther-mophillic - requiring temperatures above 40°C. The last two (i.e. b and c) do not concern fruits and vegetables for the fresh market, but may be present in inadequatelt processed items. Generally, refrigeration inhibits growth of microorganisms, but psy-chrotrophics may develop on produce if storage time is too long.

The atmosphere at which a product is stored also influences microbial development. *Clostridium botulinum*, for example, is not a concern when a product is prepared for the fresh market. However, it may develop and produce toxins on tissues with a pH higher than 4,6 and under conditions of low oxygen. It may be present in inadequately pasteurised canned products but it may also develop under modified atmosphere conditions. It has been reported that this bacterium (Table 1) can cause human poisoning.

Sale

Fruits and vegetables may become contaminated at the point of sale, storage, and home preparation. The previous discussions on personal hygiene and prevention of contact with animals are also valid here. The common practice at retail of cutting large fruits into portions (pumpkins, watermelons, melons, etc.) should be avoided and refrigeration is recommended for the most perishable items.

Fruits and vegetables are microbiologically safer compared to meat, milk, poultry, and other foods. However, during the process of fighting organisms they do not usually undergo a kill step so many organisms are not destroyed (by cooking) and are therefore potentially dangerous if contamination exists. It is difficult to estimate how real this potential threat is. This is because it is usually reported when it becomes serious. In addition as fruits and vegetables are regarded as "healthy food", they are not considered to be the cause of food-related illnesses.

Instead, other food eaten during the course of the same day is usually to blame. However, available evidence seems to indicate that this is increasingly becoming a problem. The reasons for this may be twofold: First, there is a trend towards environmentally friendly practices in agriculture. The use of manure based organic fertilisers or soil amendments increases the risk of contamination. Second, the concentration of supply particularly through supermarket distribution centers supplying a large number of stores, means that a single case of contamination can have an enormous impact throughout the system.

To obtain high quality produce with minimum risk levels, the first step is to understand the complexity of microbial contamination and recognise its importance. The examples of good agricultural and manufacturing practices described in these paragraphs may not apply to all fruits and vegetables. However, they may be useful to devise specific preventative measures. With the current level of technology, it is not possible to eliminate this risk.

However, it is important to know how to minimise it as much as possible. It is cheaper and more effective to prevent microbial contamination of fruits and vegetables than to have to confront the problem when it occurs. A successful food safety program requires serious commitment from everyone throughout the food chain from production to consumption. Key factors to be considered are: the availability of trained personnel, and a system that ensures that there are no missing links in the quality chain as regards contamination prevention.

Many different testing procedures exist for the detection of microorganisms, such as total plate counts (Figure 1) or aerobic plate count. They give an idea of the extent of microbial contamination, but have limited value in terms of assessing food safety. A wide range of microorganisms exist naturally on the surfaces of fruits and vegetables and they will colonise a growing medium. However, this does not mean that they are a health hazard. These types of tests are useful for monitoring the hygiene system or evaluating the impact of certain sanitary measures.

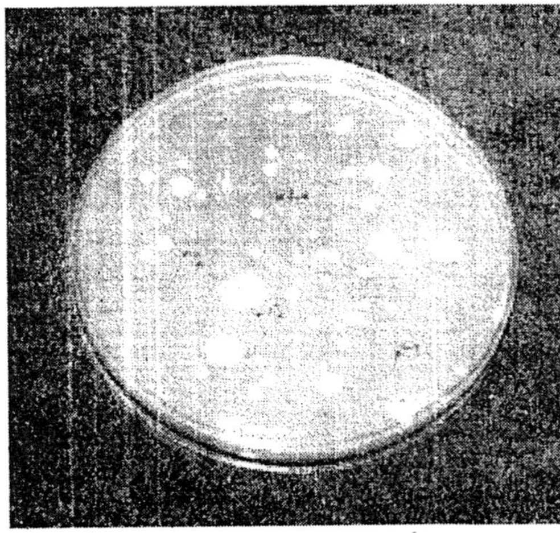

Figure 2. The total plate count of aerobic colonies on a growing media gives an idea of the degree of contamination.

Detection of *Salmonella* spp., fecal coliforms, *E. coli,* and other pathogens require specific tests and their lack of detection does not mean that the produce is free from other harmful microorganisms. As a result, the best strategy is to minimise the risk and to prevent as much contamination as possible. An important aspect in any program of good agricultural or manufacturing practices is to have a tracing system. This is because it is possible to identify and pinpoint quickly problems of contamination. The necessary corrective measures can be undertaken as soon as possible. The short lead time between harvest and consumption of fruits and vegetables makes it difficult to react in time if an outbreak is detected.

9

Selling Fruits and Vegetables

Buyer motivations are quite complex and vary according to gender, age, cultural, ethnic, regional etc. There are three different types of group behaviour patterns. The first group comprises consumers with a basic attitude. They are traditional—i.e. consumers of generic and undifferentiated fruits and vegetables. The second group seeks quality differentiation—i.e. organic or quality certified products, commercial brands, labels of certification of origin or regional produce that is differentiated etc. Convenience consumers belong to the third group. They are looking for fast and simple ways to prepare meals—i.e. prepackaged items, fresh-cut, frozen, canned and ready to eat produce.

There are other factors which also influence buying decisions. The main objective of buying is to obtain satisfaction. For fruits and vegetables, this means being able to meet nutritional requirements as well as being able to enjoy different tastes, textures, colours and aromas. There are two key considerations. The tangible quality attributes such as uniformity, freshness, quality, colour, ripeness, packaging, etc which affect appearance and make produce more appealing or attractive compared to similar products. Buying decisions are also influenced by some intangible quality attributes such as quality, environmentally friendly production techniques, brand reputation, image of the supplier, etc.

Distribution of produce to consumers can be undertaken in two ways: indirectly or directly. In the former, intermediaries (i.e. retailers, wholesalers, brokers, etc) are responsible for conveying consumer preferences to producers. The farmer prepares product to satisfy this demand. On the other hand, selling directly provides farmers with the opportunity to explore the complex range of consumer behavior and to innovate by looking for new alternatives.

Non-direct Marketing

The fruit and vegetable sector comprises many small-scale farmers on small plots in many production areas around the country. They are often located in areas distant from their main markets. This is the main reason why produce is distributed indirectly to consumers through middlemen and markets. Different commercial agreements and relationships exist between buyers and sellers. Price normally depends on volume and the quality of produce supplied.

Terminal wholesale markets are probably the most common type of marketing channel. Produce supplied from different growing areas is assembled and sold through intermediaries (wholesalers, distributors, importers, etc.) to retailers, food service companies, supermarket chains or smaller regional markets. The main advantages of terminal markets are: high concentration of supply and demand and larger volumes that can be traded. Other benefits include the fixing of reference prices for the produce traded. Fruits and vegetables should be packaged according to market handling and transport methods. In many cases palletisation is required. Wholesalers usually take ownership of produce or they can sell produce on a commission basis.

Producers located close to wholesale markets may rent space on a daily basis to sell their produce. The main benefit is the high concentration of buyers. However, because of small volumes of product for sale, they often lack bargaining power. Other marketing alternatives to selling in wholesale markets include: sales to collectors, truck drivers, shippers, packers, agents, etc. Sales locally to retail outlets (including supermarkets) provide another alternative. Purchasing directly from producers rather than through wholesalers provides some additional benefits. These include freshness, price or offering a special product that is exclusive. However, high volumes required by supermarket chains may exclude small-scale farmers as suppliers.

Other methods to gain access to large markets is by collecting product from several producers through a cooperative or selling to individual pack houses. Benefits include uniformity of quality and packaging, reduced costs and the opportunity to hire marketing specialists to increase sales and profitability.

Direct Marketing

This is sales by the farmer direct to the consumer. Different studies show that many consumers prefer direct contact with the producer/seller compared to an impersonal service, although the latter are in some cases more efficient. One of the main advantages of direct sales to consumers is the opportunity to reduce marketing costs and to add value to the product. In this way, the profit margin is increased. Producers need to become aware of existing marketing tools in order to maximise sales.

Retail Outlet

In most cities municipal ordinances regulate places and areas where fruit and vegetable retail outlets can operate. In selecting a location, the three main factors to consider are: good visibility, accessibility and proximity to buyers. Street or road crossings, the proximity of shopping centers or any other area which has the potential for high volume of passenger traffic are good locations for produce sales outlets. Some municipalities give permission to place exhibits on sidewalks to attract customers provided they do not interfere with normal pedestrian traffic.

Municipal regulations

Municipalities are autonomous and can make their own regulations for the location and operation of fruit and vegetable retail outlets.

Layout an organisation of a produce outlet

Although many customers have a written or mental list of fruits and vegetables they intend to buy, most buying decisions are made inside the store. The layout and organisation of the retail store may help customers to make a purchase thereby increasing sales. The self-service system and the traditional personal service are the two main types of marketing methods. In many cases, a combination of both are offered.

Many customers prefer the traditional system because personal interaction increases buyer confidence. Loyalty can also be built provided good quality, freshness and reasonable prices are combined with good service and friendliness. The image presented by sales staff is important to the customer. This is because they tend to think that people who take care of themselves also take an interest in the produce sold in the store. Sales staff also need to be courteous and friendly towards the customer. There are several drawbacks to traditional personal selling. First, it is not appropriate for customers who are in a hurry. In addition to this, some sales are lost because serving customers requires additional time.

The self-service system requires an attractive display of goods and a good plan for space allocation of the items on sale. This is important because produce that is not visible or attractively presented, is hard to sell. Information about varieties and prices should be clearly legible. Customers should be able to weigh produce or select prepackaged products that have been pre-weighed and labeled. This marketing method is ideal for people who prefer rapid service and prefer to choose size, ripeness, quantity and quality according to their own purchasing criteria.

The main factors to consider for increasing sales in a self-service outlet are: accessibility, visibility and easy flow of circulation. Accessibility is a physical and psychological concept. If produce is piled up high, displayed in an untidy way or difficult to reach (Figure 1), this may have a negative impact on sales. Consumers also become confused and lose time looking for goods. Ease of circulation makes shopping more convenient, particularly if trolleys can be used. As previously mentioned, visibility is a key factor in determining the whether a product can be sold.

Merchandising techniques are important to increase the visibility of the product. From a distance of 2 meters, an average person sees an area starting at 0.80 meters from the floor and up to 2 meters high and about 2-3 meters wide. Visibility decreases dramatically outside the indicated area. A minimum area of 0.30 m wide per item is required for good visibility (Figure 2).

Figure 1. A disorganised and hard to reach display confuses customers losing their time looking for the goods.

Large products like watermelons, melons, pumpkins, etc. require more space. Special allocation and large displays like islands in the middle of the aisles can be used to draw attention for quick sales (Figure 3). Slanted shelves (30–45°) and mirrors can be used to enhance product presentation (Figure 4). Refrigerated shelves should be used for highly perishable crops (Figure 5).

Strategies for maximising sales

As a rule and priority, a wide range of products should be on offer. As mentioned earlier, this is because most buying decisions take place in the store. A relatively well-supplied produce store should carry a minimum range of around 20 fruit and 30 vegetables. Product choice is not only about the range of crops on offer, but also different varieties, colours types of packaging, etc.

Figure 2. Each item for sale occupies shelf space and this depends on their individual size and marketing strategy.

Figure 3. Large items need more shelf space. Special allocation and large displays like islands in the middle of the aisles can be used to draw attention for quick sales.

Although there are no fixed rules, the proportion of fruits and vegetables on offer should be more or less equal. Vegetables during the summer months should be increased while the opposite should occur during the winter.

The quantity and type of fruits and vegetables for sale varies in each country. However, as a general rule, produce can be divided into two groups. "Basic" refers to bulk produce sales and is demanded by all types of consumers. "Specific" refers to those destined for certain niche markets.

Figure 4. Mirrors enhance products increasing their visibility

Basic products can be divided into permanent—produce that should be available on shelves all year round such as apples, tomato, potato, lettuce, carrots, etc.; seasonal, available only during certain months of the year such as peach, nectarines, melons, etc. and minor produce, such as garlic, parsley, radish, etc. Within specific categories of products are exotics—these are mainly of tropical origin and include pineapple, mango, coconut, etc.; off-season crop, in many cases originating from other countries; mushrooms; ready made salads; aromatic herbs; those of a specific quality, such as quality certified products, labeled with origin certification or regional differentiated produce, etc.; organics and fresh-cut or ready to eat products.

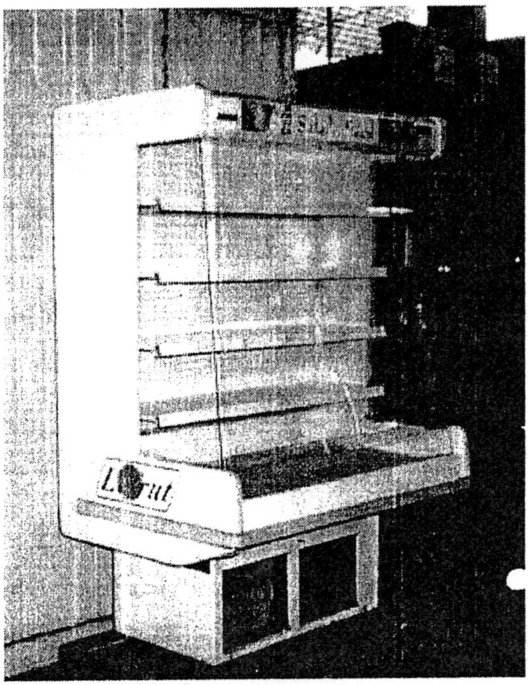

Figure 5. Refrigerated displays should be reserved for highly perishable items.

There are many different ways in which produce can be displayed and some may be highly effective. The most common practice is to place contrasting colours next to one another. This is in order to create a contrast of different coloured commodities. For example, red tomatoes next to green cucumbers, or violet and white eggplants, etc.

Figure 6. A strategy to increase sales is to group together similar products, such as tubers and roots.

Another method includes mixing and matching products that are often sold together such as tomato and lettuce, for salads, bananas with other fruits, for making fruit salads, etc. Less common is the grouping of similar products such as tubers and roots (Figure 6).

Street Selling

Although this method of marketing is frequently seen in developing countries, street selling and peddling is generally not allowed by most municipalities. There are two main reasons for this. Firstly, there are public health and hygiene considerations. As an activity it generates off-odors and insect and rodent proliferation. The second reason is that it constitutes unfair competition for established outlets. These are periodically inspected and are liable to taxes on their operations.

Ambulatory selling is undertaken in vehicles either drawn by motor, animal power or humans and produce is peddled from home to home. Street selling has the same characteristics and limitations as ambulatory selling. As scales are unavailable, produce is generally sold by units.

Community Markets

Farmers markets

A farmers market is a form of direct marketing that is located in or within proximity of a community where growers sell directly to numerous customers. Cash sales and the possibility of selling under or oversize units that cannot be marketed through other marketing channels are the main benefits of this system for farmers. For consumers it provides the opportunity to buy fresh produce or home made products and to interact with producers in an informal environment. A farmers market becomes successful when there is cooperation and interaction among three key groups:

- The sponsoring, organising or promoting group—this may be a municipality, a group of neighbors, the local Chamber of Commerce, a farmers organisation or any other association or organised group.
- Vendors, should be true farmers. They should include backyard producers. This provides a means for them to increase their income.
- Buyers, who support the market activities. It is estimated that one vendor can be supported by 800 potential buyers. So, a community of 8,000 residents could sustain a farmers market with 10 vendors.

Location of the market is an important factor. Different studies indicate that it is more practical to be located within close proximity to buyers than to vendors. Markets are often located in the town square or any other public open space. Vendors provide their own tables, racks, covers for shelter and other facilities for selling that can easily be dismantled when trading is over. Paved drives and walkways provide certain advantages in addition to adequate parking space. A tree-shaded space protected from the weather is much more desirable for both vendors and buyers.

Sponsoring or organising institutions should charge vendors a stall fee. This is in addition to providing security measures, lighting and cleaning services. They have responsibility for policy making. This includes types of persons permitted to sell at the market, fees, hours, days and months for market operation, sanctions and other operational issues. They also need to arbitrate in the event of problems and disputes. Other responsibilities include: promoting the market, avoiding conflicts through local ordinances and maintaining the market "environment". These are exactly the conditions that make a farmers market so attractive.

Fruits, vegetables, honey, eggs, firewood, flowers, plants, gardening materials and other products can be sold at farmers' markets. Bakery products, jams, marmalades, milk, homemade cheeses, and other products may require special permits. Sales of meat and other products may be forbidden. Only farm products are permitted for sale. Reselling is not allowed. Sales of crafts should be allowed because it attracts people. However, this should be carried out on a limited scale so as not to lose the spirit of this type of market.

The main advantages of selling at farmers markets include: minimum investment required to operate, no need for packaging materials, large volume of produce or a wide variety of products. Disadvantages include: low volume of unit sales, the need for customers to be served and bad weather can discourage them from attending the market.

Regional Markets

Regional markets exist in many developing countries where buyers and sellers meet to trade. From an organisational point of view they are very similar to farmers markets. One of the main differences is that operations are more concerned with wholesaling, although some retailing is undertaken. A sponsoring organisation also exists. Responsibilities include undertaking administrative duties of the market, one or more days per week for operating, stall rental on a daily basis, etc. This system provides many small-scale farmers with the opportunity to sell their produce at a fair price.

Farm Stall Sales

Farm outlets attract many customers. This form of direct marketing has the advantage of adding value. Location of the farm outlet is extremely important because it has to be seen from a certain distance. It should be located on relatively busy roads, but traffic should not be over 70-75 km/hour. The main access routes to cities are probably the best places for these types of markets. However, they can also be located in other areas such as tourist areas. Safe paved drives and availability of good parking space are factors to be considered.

Signs should clearly direct customers to the farm outlet with instructions of how to safely turn into the market. These should be seen from a certain distance and well in advance, between 100 and 2 000 m distance. This is in order to allow drivers to reduce their speed for exiting. The faster the traffic, the longer the distance of the signs from the exit, the larger the size of the letters and the lower the number of words that should be placed on the sign. For example, to be legible from 100 m, letters should be at least 30 cm high and 6 cm wide. Twenty-two letters of that size can be read at a speed of 45 km/h but only 10 at a speed of 90 km/h.

There is no standard formula for designing a farm outlet, as shelters, barns or special buildings may be used. They should be clean and tidy with enough space for displaying produce. They should have a rustic and simple appearance because this is what mainly differentiates them from other produce outlets. Preferably products produced on the farm should be available for sale. This can however be mixed with other products purchased from wholesalers.

A special type of farm sales is the "U-pick" or "pick-your-own" system. Consumers can harvest fruits and vegetables on their own. In these type of farm outlets, some produce has already been harvested and packaged. This is included with some crops that the customers harvest. Containers and harvesting tools should be available as well as precise instructions as to which areas are ready for harvesting. Sales are carried out in weight, volume, or units.

The main benefit to the farmer with this form of direct marketing method is there is no need to harvest. It also eliminates the need for sorting and packaging costs. This results in lower prices, making the produce more attractive to the consumer. The customer also has the opportunity of spending a day outdoors in contact with nature, and produce is harvested at optimum ripeness. Frequently this is a recreational event for families where the objective is to make homemade jam or marmalade. For this reason, individual sales are larger than in other direct marketing methods.

Some supervision is required here. This is because many customers do not have farm experience and may unwittingly damage plants. Moreover, the liability is higher—a higher risk of accidents can occur with harvesting tools, ladders or equipment. A good emergency system should be provided including an insurance policy. Fruit crops are more appropriate than vegetables and different varieties and long harvesting periods are the ideal crops for this system. Pesticide applications and waiting periods should be carefully planned. This is so that there is always an area ready to be harvested.

The basic rule for maximising farm sales is the concept: "the longer they stay, the more they will spend". For this reason, additional programs should be offered such as farm tours, wagon rides, activities and games for kids, on-farm walking trails, camping, craft demonstrations, nature study, farm animal petting zoo, fishing, etc. Other ways of increasing income is through sales of homemade jams and marmalades, sauces, traditional or special recipes, homemade food, crafts, etc.

Selling to Restaurants and Hotels

Direct buying by restaurants, hotels, hospitals, geriatric institutions, etc. is undertaken to reduce costs and to simplify the daily supply and preparation of different dishes served. Steady all year round demand is the main advantage to the farmer. The farmer has the opportunity of adding value to the product by washing, peeling, pitting, slicing, portioning, etc. One of the main disadvantages is the difficulty of satisfying steady demand with seasonal products and prices can vary significantly. Other factors include: limited storage space in restaurants, regular delivery of small volume orders, higher telephone and transport costs and extra documentation which increase operational costs.

Ethnic and 'high class' restaurants belong to a special category and need to be taken into account. This is because they often require special or premium quality produce. Contracts can be profitable provided quality produce is supplied according to specification and timely delivery. Also, a creative chef can significantly expand sales.

10

Preparation for the Fresh Market

Marketing is important to all farm enterprises, regardless of size. Some farmers, such as cash grain farmers or dairy farmers, have large, well-established markets. They can use existing organisations to perform the marketing function for them, or they can band together, form a cooperative, and market their products jointly. Small-scale fruit and vegetable growers generally have more difficulty finding established markets; therefore, they usually develop marketing systems tailored to their own unique situations. It is strongly recommend that you identify and research your market before you become a fruit and vegetable grower.

Fruits and vegetables are produced seasonally, but the market requires products throughout the year. For many decades, this problem of matching product availability with consumer demand was solved by two methods: (1) selling fresh products during harvest and shortly thereafter, and (2) processing the rest to meet demand during the remainder of the year. As technology improved and consumer incomes increased, it became possible to provide fresh produce year-round.

After harvest, fruits and vegetables need to be prepared for sale. Regardless of the destination, preparation for the fresh market comprises four basic key operations:

1. Removal of unmarketable material,
2. Sorting by maturity and/or size,
3. Grading,
4. Packaging.

Any working arrangement that reduces handling will lead to lower costs and will assist in reducing quality losses. Market preparation is therefore preferably carried out in the field. However, this is only really possible with tender or perishable products or small volumes for nearby markets. Products need to be transported to a packinghouse or packing shed in the following cases: for large operations, distant or demanding markets or products requiring special operations like washing, brushing, waxing, controlled ripening, refrigeration, storage or any specific type of treatment or packaging.

These two systems (field vs. packinghouse preparation) are not mutually exclusive. In many cases part field preparation is completed later in the packing shed. Because it is a waste of time and money to handle unmarketable units, primary selection of fruits and vegetables is always carried out in the field. In this way products with severe defects, injuries or diseases are removed. Lettuce is an example of field preparation where a team of three workers cut, prepare and pack.

For distant markets, boxes prepared in the field are delivered to packhouses for palletizing, precooling, and sometimes cold storage before shipping. Mobile packing sheds provide an alternative for handling large volumes in limited time. Harvest crews feed a mobile grading and packing line. On completion of loading, the consignment is shipped to the destination market and replaced by an empty truck. In mechanised harvesting, the product is transported to the packhouse where it is prepared for the market. In many cases, harvest crews make use of an inspection line for primary selection on the field.

Packinghouses

A packinghouse allows special operations to be performed. Another advantage (over field preparation) is that products can be prepared continuously for 24 hours regardless of the weather. With its capacity to process large volumes, farmers associations, cooperatives, or even community organisations can take advantage of these opportunities. The size and degree of complexity of a packing shed depends on the following factors: crop(s) and volume to be processed, capital to be invested, its objectives such as handling of owner's production or to provide service to others.

Packing sheds range from a straw shelter to highly automated facilities. In some cases storage rooms as well as offices for commercial sales are annexed to packing sheds. A packhouse can be defined as a place protected from weather for both, product and personnel. It is organised in such a way that product is prepared in a centralised handling operation. To some extent, this is similar to a factory assembly line, where raw material from the field undergoes a sequence of activities resulting in the final packaged product.

A packinghouse needs to be located close to the production area and within easy access to main roads or highways. It also needs to have one entrance to facilitate and control supply and delivery. Moreover, it needs to be large enough for future expansion or additional new facilities. Sufficient space outside is also required to avoid congestion of vehicles entering and leaving. Buildings should be designed to ensure sufficient shade during most of the day in the loading and unloading areas. They also need good ventilation in summer and protection in winter.

Packinghouses are usually built with cheap materials. However, it is important to create a comfortable environment both for produce and workers. This is because product exposed to unfavorable conditions can lead to rapid deterioration in quality. Also, uncomfortable working conditions for staff can lead to unnecessary rough handling. A packinghouse should have adequate room for easy circulation with ramps to facilitate loading and unloading.

Doors and spaces should be sufficiently large to allow the use of forklifts.The reception area should be large enough to hold product equivalent to one working day. The main reason for this is to keep the packinghouse in operation in the event of an interruption in the flow of product from the field (rain, machine breakdown, etc).

Electricity is critical for equipment, refrigeration and particularly lighting. Because packhouses usually work extended hours or even continuously during harvest time, lighting (both, intensity and quality) is critical in identifying defects on inspection tables. Lights should be below eye level to prevent glare and eyestrain. Light intensity should be around 2 000-2 500 lx for light coloured products but 4 000-5 000 for darker ones. The working area together with the whole building should have lighting. This is in order to avoid the contrasts caused by shaded areas, resulting in temporary blindness when the eyes are raised.

Dull colours and non-glossy surfaces are a requirement for equipment, conveyor belts and outfits. In this way, defects are not masked because of the reflection of light. It also helps to reduce eye fatigue. A good supply of water is important for washing product, trucks, bins and equipment, as well as for dumping. In some cases it may also be necessary for hydro cooling. Provision of an adequate waste water disposal system is as important as a good source.

Administration offices should be located on clean and quiet areas and if possible elevated. This is so that the entire operation is visible. Packinghouses should have facilities or laboratories for quality analysis. After working out the details of the building layout, it is important to prepare a diagram for the movement of product throughout the packinghouse and activities to be undertaken for the entire operations. Handling must be minimised and movement of product should always be in one direction without crossovers. It may be possible to undertake operations concurrently, such as working simultaneously on different sizes or maturity stages.

Preparation and Packing Operations

Preparation and packing operations should be designed to minimise the time between harvest and delivery of the packaged product. Reception is one area where delays frequently occur and the product should be protected from the sun as much as possible. Product is normally weighed or counted before entering the plant and in some cases samples for quality analysis are taken. Records should be kept, particularly when providing a service to other producers. Preparation for the fresh market starts with dumping onto packinghouse feeding lines.

Dumping may be dry or in water. In both cases it is important to have drop decelerators to minimise injury as well as control the flow of product. Water dipping produces less bruising and can be used to move free-floating fruits. However, not all products tolerate wetting. A product with a specific density lower than water will float, but with other products salts (sodium sulfate, for example) are diluted in the water to improve floatation.

Water dipping through washing helps to remove most dirt from the field. For thorough cleaning, more washings and brushing are required. Water rinsing allows produce to maintain cleanliness and be free of soil, pesticides, plant debris and rotting parts. However, in some cases this is not possible. This is because of insufficient water. If recirculated water is used, this needs to be filtered and settled dirt removed.

Chlorination of dumping and washing waters with a concentration 50-200 ppm of active chlorine, eliminates fungi spores and bacteria on the surface of diseased fruits. This prevents the contamination of healthy fruit. In addition to this, bruising should be avoided since this is the entry for infection by decay organisms. At depths greater than 30 cm and for periods of time longer than 3 minutes, water tends to penetrate inside fruits, particularly those that are hollow such as peppers. Water temperature also contributes to infiltration. It is recommended that fruit temperature is at least 5°C lower than liquid.

After dumping, the first operation that usually follows is the removal of unmarketable material. This is because handling of plant material that cannot be sold is costly. This is performed prior to sizing and grading. Primary selection is one of the four basic operations for market preparation carried out in the field. This step involves the removal of over mature, too small, severely damaged, deformed or rotting units.

Very small produce is usually mechanically removed by mesh screens, pre-sizing belts or chains. Bruised, rotted, off-shaped units, wilted or yellow leaves are usually removed by hand. Garlic and onions are topped to remove the dry foliage attached to the bulbs by specific equipment and in many crops soil and loose parts are removed by brushing.

In crops where water dipping is possible, differential floatation could be used to separate rejects. In addition to this, detergents and brushes can be used to remove soil, latex, insects, pesticides etc. Clean fruits should be dried with sponges or hot air. Culls as well as other plant parts from cutting, peeling, trimming, bruised and spoiled fruits can be used for animal feeding. Although they provide a good source of energy and are extremely tasty, their high water content makes them bulky and expensive to transport.

In addition to this, their nutritional value is less than other food sources. This is because of their low protein and dry matter contents (in terms of volume). Their inclusion in the diet must be in the right proportions to avoid digestive problems. Another disadvantage is that in many cases they are highly perishable and cannot be stored.. This means that they cannot be gradually introduced into the animal's diet. When not used for animal feeding, they can be disposed as sanitary fillings or organic soil amendments.

Sizing is another basic operation undertaken in a packhouse and can be carried out before or after sorting by colour. Both operations should always be carried out before grading. This is because it is easier to identify units with defects on a uniform product, either in terms of size or colour. There are two basic systems—according to weight or dimensions (diameter, length or both).

Spherical or almost spherical products like grapefruits, oranges, onions, and others, are probably the easiest to sort by size. Several mechanisms are available from mesh screens to diverging belts or rollers with increased spaces between them. Sizing can also be performed manually using rings of known diameter. Sorting by weight is carried out in many crops with weight sensitive trays. These automatically move fruit onto another belt aggregating all units of the same mass.

Amongst the four basic operations, grading is probably the most important. It consists of sorting product in grades or categories of quality. Two main systems exist: static and dynamic. Static systems are common in tender and/or high value crops. Here the product is placed on an inspection table where sorters remove units which do not meet the requirements for the grade or quality category.

The dynamic system is probably much more common. Here product moves along a belt in front of the sorters who remove units with defects. Main flow is the highest quality grade. Often second and third grade quality units are removed and placed onto other belts. It is much more efficient in terms of volume sorted per unit of time. However, personnel should be well trained. This is because every unit remains only a few seconds in the worker's area of vision. There are two types of common mistakes: removing good quality units from the main flow and more frequently, not removing produce of doubtful quality. Rejects mainly on aesthetic grounds provide a second or even third quality grade. These can be marketed in less demanding outlets or used as raw material for processing.

Small scale processing, however, needs to be able to achieve a standard of quality similar or even better than large industries. This is not always possible because industrial plants tend to use specific varieties and processes. In addition to this, surpluses for the fresh market and sub-standard products do not provide uniform raw material. The industrial yield is low and this together with the low technology in the manufacturing process can result in a product of variable quality. At this point, it is important to highlight that the quality of a processed product will depend both upon, the quality of the raw material and the manufacturing process.

Special Operations

These operations are commodity specific. They are different from basic operations because they are carried out on every crop independent of size and sophistication of the packinghouse.

Colour Sorting

These are common in fruits and fruit vegetables and can be undertaken electronically. Fruits are usually harvested within a range of maturity that needs to be uniform for sale. Harvesting within a narrow range of maturity reduces colour sorting. However, this is only possible for low-volume operations.

Waxing

Some fruits such as apples, cucumbers, citrus, peaches, nectarines and others, are waxed for the following reasons: to reduce dehydration, improve their postharvest life by replacing the natural waxes removed by washing and to seal small wounds produced during handling. Waxes are also used as carriers of some fungicides or just to increase shine and improve appearance.

Different types and formulae of waxes are available. These can be applied as sprays or foams, or by immersion and dripping or in other ways. Uniform distribution is important. Soft brushes, rollers or other methods are used to ensure that application on the surface of fruit is thorough and texture is even. Heavy application can block fruit gas exchange and produce tissue asphyxia. Internal darkening and development of off-flavors and off-odors are some of the characteristics. It is very important that waxes are approved for human consumption.

Degreening

The main causes of greening are climatic conditions before harvest. For example, citrus often reaches commercial maturity with traces of green colour on the epidermis (flavedo). Although not different from fruits with colour, consumers sense that they are not ripe enough and have not reached their full flavor. Degreening consists of chlorophyll degradation to allow the expression of natural pigments masked by the green colour.

In purpose built chambers, citrus fruits are exposed from 24 to 72 hours (depending on degree of greening) to an atmosphere containing ethylene (5-10 ppm) under controlled ventilation and high relative humidity (90-95%). Conditions for degreening are specific to the production area. Artés Calero (2000) recommends temperatures of 25-26 °C for oranges, 22-24 °C for grapefruit and lemon and 20-23 °C for mandarins.

Controlled Ripening

Maturity at harvest is the key factor for quality and postharvest life. When shipped to distant markets, fruits need to be harvested slightly immature (particularly climacteric ones) to reduce bruising and losses during transport. Prior to distribution and retail sales, however, it is necessary to speed up and achieve uniform ripening. The main reason for this is so that product reaches consumers at the right stage of maturity. As with degreening, ethylene is used but at higher concentrations.

Banana provides a typical example of this type of operation. It can however, also be carried out on tomatoes, melons, avocados, mangoes and other fruits (Table 1). Controlled ripening is performed in purpose built rooms where temperature and relative humidity can be controlled and ethylene removed when the process has been completed. The process involves initial heating to reach the desired pulp temperature. This is followed by an injection of ethylene at the desired concentration.

Under these conditions, the product is maintained for a certain amount of time followed by ventilation in order to remove accumulated gases. On completion of the treatment, the temperature is reduced to the desired level for transportation and/or storage. Ethylene concentration and exposure time are a function of temperature, which accelerates the process.

Table 1. Conditions for controlled ripening of some fruits.

	Ethylene concentration (ppm)	Ripening temperature °C	Exposure time to these conditions (hr.)
Avocado	10-100	15-18	12-48
Banana	100-150	15-18	24
Honeydew melon	100-150	20-25	18-24
Kiwifruit	10-100	0-20	12-24
Mango	100-150	20-22	12-24
Stone fruits	10-100	13-25	12-72
Tomato	100-150	20-25	24-48

Pest Management

Different treatments are performed to prevent and control pests and diseases at postharvest level. Fungicides belonging to different chemical groups are widely used in citrus, apples, bananas, stone fruits and other fruits. Most have a fungistatic activity. This means that they inhibit or reduce germination of spores without complete suppression of the disease. Chlorine and sulfur dioxide are amongst those most widely used.

Chlorine is probably the most widely used sanitiser. It is used in concentrations from 50 to 200 ppm in water to reduce the number of microorganisms present on the surface of the fruit. However, it does not stop the growth of a pathogen already established. Table grapes are usually fumigated with sulfur dioxide to control postharvest diseases at a concentration of 0,5% for 20 minutes followed by ventilation. During storage, periodic (every 7-10 days) fumigations are performed in concentrations of 0.25%.

During transport, pads impregnated with sodium metabisulfite can be used inside packages. These slowly generate sulfur dioxide in contact with the humidity released by fruits. Gas fumigation is the most important method for eliminating insects, either adults, eggs, larvae or pupae. Methyl bromide was probably the most widely used fumigant for many years but it is banned in most countries. It has been replaced by temperature (high and low) treatments, controlled atmospheres, other fumigants or irradiation. It is also possible to prevent some postharvest physiological disorders with chemical treatments.

For example, calcium chloride (4-6%) dips or sprays for bitter pit in apples. Other methods include dipping or drenching fruits in chemical solutions to avoid storage scalds or other disorders. Similarly, the addition of low concentrations of 2.4-D to waxes assists in keeping citrus peduncles green.

Heat Treatments

Cold can be used in low temperature tolerant fruits (apples, pears, kiwifruit, table grapes, etc.) and other potential carriers of quarantine pests and/or their ovipositions. Exposure to any of the following combinations of temperatures and time is provided in the following recommendations (Table 2). Heat treatments like hot water dips or exposure to hot air or vapor have been known for many years for insect control (and for fungi, in some cases).

When restrictions were extended to bromine based fumigants, however, heat treatments were reconsidered as quarantine treatments in fruits such as mango, papaya, citrus, bananas, carambola and vegetables like pepper, eggplant, tomato, cucumber and zucchinis. Temperature, exposure and application methods are commodity specific and must be carried out precisely in order to avoid heat injuries, particularly in highly perishable crops.

On completion of treatment, it is important to reduce temperature to recommended levels for storage and/or transport. Hot water immersion requires that fruit pulp temperature is between 43 and 46,7 °C for 35 to 90 minutes. This depends on commodity, insect to be controlled and its degree of development. Dipping in hot water also contributes to reduced microbial load in plums, peaches, papaya, cantaloupes, sweet potato and tomato but does not always guarantee good insect control. For the export of mangoes from Brazil, it is recommended that dipping is performed at 12 cm depth in water at 46,1 °C and for 70-90 minutes.

Table 2. Combinations of temperature and exposure time for fruit fly quarantine treatments.

Time (days)	Maximun temperature (°C)	
	Ceratitis capitata	Anastrepha fraterculus
10	0,0	
11	0,6	0,0
12	1.1	
13		0,6
14	1,7	
15		1,1
16	2,2	
17		1,7

Many tropical crops are exposed to hot and humid air (40-50 °C up to 8 hours) or water vapor to reach a pulp temperature which is lethal to insects. Hot air is well tolerated by mango, grapefruit, Navel oranges, carambola, persimmon and papaya. Similarly, vapor treatments have been approved by the USDA-APHIS (U.S. Department of Agriculture, Animal and Plant Health Inspection Service) for clementines, grapefruits, oranges, mango, pepper, eggplant, papaya, pineapple, tomatoes and zucchinis.

Sprout Suppression

In potatoes, garlic, onion and other crops, sprouting and root formation accelerate deterioration. They also determine the marketability of these products. This is because consumers strongly reject sprouting or rooting products. After development, bulbs, tubers and some root crops enter into a "rest" period. This is characterised by reduced physiological activity with non response to environmental conditions.

In other words, they do not sprout even when they are placed under ideal conditions of temperature and humidity. Different studies show that during rest, endogenous sprout inhibitors like abscisic acid predominate over promoters like gibberellins, auxins and others. This balance changes with the length of storage to get into a "dormant" period. They will then sprout or form roots if placed under favorable environmental conditions. There are no clear-cut boundaries between these stages.

Instead, there is a slow transition from one to the other as the balance between promoters and inhibitors change. With longer storage times, promoters predominate and sprouting takes place. Refrigeration and controlled atmospheres reduce sprouting and rooting rates but because of their costs, chemical inhibition is preferred. In onions and garlic Maleic Hydrazide is sprayed before harvest while in potatoes CIPC (3-chloroisopropyl-Nphenylcarbamate) is applied prior to storage as dust, immersion, vapor or other forms of application. As CIPC interferes with periderm formation, it must be applied after curing is completed.

Gas Treatments

Different studies have shown that exposure to carbon dioxide rich atmosphere (10-40% up to week) before storage, contributes towards maintaining quality in grapefruits, clementines, avocados, nectarines, peaches, broccoli and berries. Control of insects is possible with higher concentrations (60-100%). The effect of this gas is not well understood. What is known is that it has an inhibitory effect on metabolism and ethylene action and the effect is persistent after treatment.

Also, at higher concentrations (> 20%) there is difficulty in spore germination and growing of decay organisms. Similarly, exposure to very low oxygen atmosphere (< 1%) also contributes towards preserving quality and controlling insects in oranges, nectarines, papaya, apples, sweet potatoes, cherries and peaches. Lowering oxygen concentration reduces respiratory rate and the whole metabolism.

Purpose of Packaging

The main purpose of packaging is to ensure that the product is inside a container along with packing materials to prevent movement and to cushion the produce (plastic or moulded pulp trays, inserts, cushioning pads, etc.) and for protection (plastic films, waxed liners, etc.). It needs to satisfy three basic objectives. These are to:

1. Contain product and facilitate handling and marketing by standardising the number of units or weight inside the package.
2. Protect product from injuries (impact, compression, abrasion and wounds) and adverse environmental conditions (temperature, relative humidity) during transport, storage and marketing.
3. Provide information to buyers, such as variety, weight, number of units, selection or quality grade, producer's name, country, area of origin, etc. Recipes are frequently included such as nutritional value, bar codes or any other relevant information on traceability.

A well-designed package needs to be adapted to the conditions or specific treatments required to be undertaken on the product. For example, if hydrocooling or ice-cooling need to be undertaken, it needs to be able to tolerate wetting without losing strength; if product has a high respiratory rate, the packaging should have sufficiently large openings to allow good gas exchange; if produce dehydrates easily, the packaging should provide a good barrier against water loss, etc. Semi-permeable materials make it possible for special atmospheres inside packages to be generated. This assists in maintaining produce freshness. There are three types of packaging:

1. Consumer units or prepackaging
2. Transport packaging
3. Unit load packaging or pallets

When weighed product reaches the consumer in the same type of container in which it is prepared—this is described as a consumer unit or prepackaging. Normally, this contains the quantity a family consumes during a certain period of time (300 g to 1,5 Kg, depending of product). Materials normally used include moulded pulp or expanded polystyrene trays wrapped in shrinkable plastic films, plastic or paper bags, clamshells, thermoformed PVC trays, etc.

Onions, potatoes, sweet potatoes etc are marketed in mesh bags of 3-5 Kg. Colours, shapes and textures of packaging materials play a role in improving appearance and attractiveness. Transport or packaging for marketing usually consists of fiberboard or wooden boxes weighing from 5 to 20 Kg or bags can be even heavier. They need to satisfy the following requirements: be easy to handle, stackable by one person; have the appropriate dimensions so that they fit into transport vehicles and materials should be constructed with biodegradable, non-contaminating and recyclable materials.

Packaging intended for repeated use should be: easy to clean and dismantle so that it is possible to significantly reduce volume on the return trip; ability to withstand the weight and handling conditions they were designed for, and meet the weight specifications or count without overfilling. In these type of packages it is common to use packaging materials which serve as dividers and immobilise the fruit.

For example, vertical inserts can be used. They also assist in reinforcing the strength of the container, particularly when large or heavy units such as melons or watermelons are packed. Trays also have the same objective but they separate produce in layers. They are common in apples, peaches, plums, nectarines, etc. Plastic foam nets are used for the individual protection of large fruits like watermelons, mango, papayas, etc. It is also possible to use paper or wood wool, papers or other loose-fill materials.

In many developing countries containers made of natural fiber are still used for the packaging of fruits and vegetables. Although cheap, they cannot be cleaned or disinfected. They therefore represent a source of contamination of microorganisms when reused. Moreover, there is a risk of bruising as a result of compression. This is because they were not designed for stacking. In addition to this, the significant variations in weight and/volume makes marketing a complex business.

Finally, pallets have become the main unit load of packaging at both domestic and international level. Their dimensions correspond to those of maritime containers, trucks, forklifts, storage facilities, etc. As unit loads they reduce handling in all the steps in the distribution chain. Different sizes exist. However, the most common size internationally is 120 x 100 cm. It is sometimes made of plastic materials.

Depending on the packaging dimensions, a pallet may hold from 20 to 100 units. To ensure stability, pallet loads are secured with wide mesh plastic tension netting or a combination of corner post protectors and horizontal and vertical plastic strapping. In many cases individual packages are glued to each other with low tensile strength glue that allow separate units but prevent sliding. They are also stacked crosswise or interlocked to contribute to the load stability.

There is a trend towards standardisation of sizes. This is because of the wide variety of shapes and sizes of packaging for fruits and vegetables,. The main purpose of standardisation is to maximise utilisation of the pallet's surface based on the standard size 120 x 100 cm. The ISO (International Standards Organisation) module (norm ISO 3394) sets 60 and 40 cm as basic horizontal dimensions divided in subunits of 40 x 30 cm and 30 x 20 cm.

There are no regulations regarding the height of individual packages. However, the palletised load should not exceed 2.05 m to ensure safe handling. On the recommendation of USDA, the MUM system (Modularisation, Unitisation and Metrication) also has as its objective, container standardisation on the basis of the 120 x 100 cm pallet.

The trend towards the use of non-returnable containers poses an environmental challenge. To reduce the impact, packages need to be designed to meet their functional objectives, with minimal wastage of materials and need to be recyclable, after their main functional use.

Bibliography

Ashby, B.H., R.T. Hinsch, L.A. Risse, W.G. Kindya, W.L. Craig; and M.T. Turezyn, *Protecting Perishable Foods During Transport by Truck*, HBK-669. USDA, AMS, September, 1987.

Cassell, Clark C., *The World is Your Market - An Export Guide for Small Business*, Edited by William Delphos, Braddock Communications, Washington, DC, 1990.

CIGNA Companies, *Ports of the World: a guide to cargo loss control*, 14th ed. Philadelphia, PA.

Fuell, L. D., D. C. Miller, and M. Chesley, *Dictionary of International Agricultural Trade*, USDA, FAS, June 1988.

Friedheim, Amy, *Ocean Liner Freight Rates and Services*, USDA, AMS, May 1991.

Hardenburg, R. E., A. E. Watada, C. Y. Wang, *The Commercial Storage of Fruits, Vegetables, and Florist and Nursery Stocks*, HBK-66. USDA, ARS, September 1986.

Lipton, W.J. and J.M. Harvey, *Compatibility of Fruits and Vegetables During Transport in Mixed Loads*, MRR-1070. USDA, Fresno, CA, 1977.

McGregor, Brian M., *Tropical Products Transport Handbook*, HBK-668. USDA, AMS, September 1989.

Nicholas, C. J., *Export Handbook for U.S. Agricultural Products*, HBK-593. USDA, AMS, March 1985.

Safeway Stores, Inc. *Transit and Storage Properties of Produce Commodities*. Produce Marketing Association, Newark, DE. 1986.

The Bureau of National Affairs, Inc. *International Trade Reporter Export Reference Manual*. Washington, DC, 1991.

U.S. Department of Agriculture, Foreign Agricultural Service, *Food & Agricultural Export Directory 1991*. MPN-1497, September 1991.

U.S. Department of Commerce, Bureau of the Census, *Correct Way to Fill Out the Shipper's Export Declarations*, January 1988.

U.S. Department of Commerce, Federal Express Corporation, *A Basic Guide to Exporting*, January 1992.

Index

Abscission layer 9
Agrochemicals 219
Air circulation 94
Air recirculation 24
Air transportation 105
Ascorbic acid 152
Atmosphere-controlled environments 23

Bulk bins 21
Controlled Atmosphere (CA) 87
Cargo lifting equipment 101
Chilling injury 34
Chlorination 263
Climacteric fruits, 12

Dehydration 207
Dehydrator Drying 148
Disease control programme 222
Distribution chain 275
Dried vegetables 147

Enclosed Vehicle Transportation 89

Fibreboard boxes 29
Food and Drug Administration (FDA) 154
Fruit Firmness testers 9

Galvanized screening 148
Generally Recognised as Safe (GRAS) 154

Hazard Analysis Critical Control Points (HACCP) 225
Highly perishable vegetables 33
Horticultural crop 79

Inland transportation 87
Inter-island trade 102
Internal quality 204

Low-cost technology farming systems 1
Lower-grade products 80

Manufacturing process 266
Marketing channel 82
Maturity Index 3
Maximum Residue Levels (MRLs) 220
Mechanical damage 1
Microwave Method 162
Modernization 1

Night ventilation 36
Non-biotic contaminants 206

Off-shaped units 264
Open Vehicle Transportation 89
Overlaps maturation 6

Pallet boxes 30
Pallet-trucks 85
Parasitic diseases 82

Penetrometer 9
Pest management 269
Physiological deterioration 81
Post-harvest handling 22, 79
Post-harvest processing 22
Production-commercialisation-consumption chain. 222

Quality assurance 79

Reefer containers 104
Reefer vessels 103
Refrigerated Sea Transport 103
Refrigerated Vehicle Transportation 92
Roller-conveyors 85

Salt Solution Dip 153
Small scale processing 265
Sodium-Ortho-Phenyl-Phenate (SOPP) 26
Storage Temperature 33
Sun Drying 149
Syrup Blanching 153

Temperature management 88
Texturometer 9
Trans-shipment 107
Truck-based mechanical refrigeration systems 92

Uniformity 207
Unit Load Containers (ULD) 106

Water Blanching 155
Water dipping 262